"十四五"普通高等教育本科部委级规划教材

服装导论
（双语）

戴晓群　卢业虎　主编

洪　岩　副主编

中国纺织出版社有限公司

内 容 提 要

本书立足于全球化服装市场背景，系统介绍了服装全产业链的各个环节。在内容设置上，以产品开发流程为主线，分为商业策划、创意策划、技术和生产规划、分销计划四大模块。通过各章内容的呈现，使学生逐步熟悉从商业计划、流行趋势预测、面料开发、产品系列设计、技术规格制订、大货生产到批发和零售计划的时装产品开发全过程。通过对国际化行业情境的设置，使学生系统学习本专业的理论知识与专业英语表达。

本书可作为服装专业"服装导论"或"时装产品开发"及相关课程的教材；也可作为纺织类专业或国际贸易专业选修课程的教材。

图书在版编目（CIP）数据

服装导论：双语：汉文、英文 / 戴晓群，卢业虎主编；洪岩副主编. --北京：中国纺织出版社有限公司，2022.6

"十四五"普通高等教育本科部委级规划教材

ISBN 978-7-5180-9292-5

Ⅰ. ①服… Ⅱ. ①戴… ②卢… ③洪… Ⅲ. ①服装—理论—高等学校—教材—汉、英 Ⅳ. ①TS941

中国版本图书馆CIP数据核字（2022）第005168号

责任编辑：沈　靖　孔会云　　责任校对：寇晨晨　　责任印制：何　建

中国纺织出版社有限公司出版发行
地址：北京市朝阳区百子湾东里A407号楼　邮政编码：100124
销售电话：010—67004422　传真：010—87155801
http://www.c-textilep.com
中国纺织出版社天猫旗舰店
官方微博 http://weibo.com/2119887771
唐山玺诚印务有限公司印刷　各地新华书店经销
2022年6月第1版第1次印刷
开本：787×1092　1/16　印张：12.25
字数：223千字　定价：58.00元

前言

　　在工业4.0背景下，随着人工智能、元宇宙、智能可穿戴等新兴技术的推广应用，服装全球化加剧，消费者市场不断演变，时装业呈现出全新的面貌，对服装专业学生的创新能力需求日益突出。我国的高等教育也进入高质量发展、内涵发展的新阶段，高校的服装专业教学应该做出符合行业发展和高等教育发展的变革。"服装导论"课程是服装专业学生在正式进入专业学习之前初次接触的专业课程，是对整个服装学科和服装业的简介，更是塑造对本专业全工程系统思维、培养对专业学习热情的关键课程。服装市场和供应链的全球化加剧，使得双语和全英文教学对服装专业学生有着独特的必要性。在此背景下，编者团队立足当下需求，编纂了本教材。

　　考虑到21世纪大学生英文水平普遍提高，本教材摒弃了中英双语对照的形式，同时全文使用尽可能简洁明了的英文，避免使用生僻的英文词汇。本教材内容有以下几方面特色和创新：

　　（1）内容设计上：在全球化服装市场背景下，使学生置身国际化行业情境中学习本专业的理论知识与专业英语词汇，兼具理论体系性与行业实践性。

　　（2）框架设计上：为帮助学生形成专业的全工程系统认识，以产品开发流程为主线，分为商业策划、创意策划、技术和生产规划、分销计划四大模块，把全球化服装供应链上各环节有机串联起来，内容连贯。

　　（3）编排思路上：为活跃课堂教学和课后自主学习，每章附有丰富的课堂和课后活动设计，并有附录和相关网页等扩充资料作为教材的开放式补充，提供学生深入学习提高的拓展资源。

　　（4）课外实训上：为强化实践训练，设计了"Semester project"产品开发项目，让学生逐步完成整个产品开发实践，可作为课程学习效果的综合评估，充分体现育人于过程的思想，加强过程化学习效果评估，提升学生的创新能力。

　　基于以上特色，本教材既可作为服装专业低年级的专业导论课教材，也可作为本专业高年级的时装产品开发专业课教材，同时可作为纺织类专业或国际贸易专业选修课程的教材。

　　全书共13章，第一、二、三、十一、十二章由戴晓群编写；第四~八章由洪岩编写；第九、十、十三章由卢业虎编写，戴晓群负责全书的统稿工作。研究生宁晨辰、冯胜楠、高珊，本科生刘佳媛、宋雨晴、樊晨星、王一磊协助完成了相关图文资料的收集和整理。在此，向以上人员表示衷心的感谢。

在本书的编写和出版过程中，得到了苏州大学纺织与服装工程学院和中国纺织出版社有限公司有关领导、老师和编辑的大力支持和帮助，对此谨表示诚挚的谢意。此外，鉴于编者的专业和英语水平有限，书中不妥之处敬请指正。

编者

2021年12月

Contents

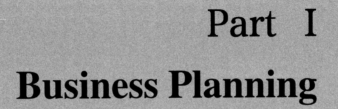

Part I

Business Planning

Chapter 1

Introduction to the Fashion Industry

课题名称： Introduction to the Fashion Industry

课题内容： 时尚及时装行业简介

课题时间： 4课时

教学目的：

1. 理解时尚和时装的概念。

2. 掌握时装行业的定义。

3. 了解时装产品开发过程。

4. 了解时装产品开发过程的动态性。

5. 理解全球化服装供应链。

课前准备： 浏览提供的相关网页，对时装行业有初步感性认识。

1.1 Fashion and products

1.1.1 What is fashion ?

Fashion is to do with change

Fashion is a popular aesthetic expression in a certain time and context, especially in clothing, footwear, lifestyle, accessories, makeup, hairstyle and body proportions. The term fashion applies to anything that's of the moment and subject to change; it's anything that members of a population deem desirable and appropriate at a given time. In this book, the concept of fashion will be taken to deal with garments and accessories. Figure 1.1 identifies some major categories of clothing along with their main usage situations, but this list is by no means exhaustive.

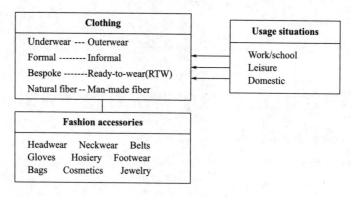

Figure 1.1 Fashion products

Ever since the origin of mankind, fashion has had the power to fascinate and excite. Fashion denotes change, something new and different from what existed before. The trend is a general direction or movement for the change. A trend is what's hip or popular at a certain point in time. A fashion trend usually refers to a certain style. In the course of history, 100 years isn't a particularly long time. But in the course of fashion history, it's the difference between trying to get around in a bone-cinching girdle and ankle-length skirt and easing into Vans sneakers you're probably wearing today. Table 1.1 lists the most influential trends from the 1900s through today.

Table 1.1 The most influential trends from the 1900s through today

Time by decades	Trends
1900s	The S-Bend Corset
1910s	The Hobble Skirt
1920s	Flapper Style
1930s	Bias-Cut Gowns
1940s	The Bikini

continued

Time by decades	Trends
1950s	The "New Look"
1960s	The Miniskirt
1970s	Platform Heels
1980s	Leggings
1990s	Minimalism
2000s	The Tracksuit
2010s	Athleisure

Fashion is about creating

In order for the change which is intrinsic to fashion to take place, the industry must continually create new products. Used in another sense, the term fashion means to construct, mould or make. Fashion, therefore, also involves a strong creative and design component.

Design is defined as "a creative process, driven by a need, which leads to an invention of some sorts, be it practical or artistic, functional or simply attractive, devised to enhance life in some way". The role of designer in the fashion industry is crucial to its success. The task is specifically one of interpreting society's current and anticipated mood into desirable, wearable, garments for every type and level of market. Design skill is essential and can be seen in all products from the made-to-measure suit to the elaborate embroidery on a cardigan. The level of design can vary considerably from a basic item to the artistic creations of Armani.

1.1.2 The nature of fashion products

Fashion products are consumer products. A consumer is an individual who wears or uses a product. Different from industrial products, consumer products are products that are bought by individuals or households for personal use.

Attributes of fashion products

In designing and developing consumer products, it is essential to understand what is being offered and therefore to appreciate the customer's perception of the product. The fashion consumer will tend to view the garment as a series of attributes. These attributes can be classified as three categories: core, tangible, and intangible.

The core attributes revolve around three basic functions of clothing: protection, modesty and adornment. Since every garment will offer combinations of these functions to a greater or lesser degree, the consumer must decide how well the combination matches his or her basic criteria for purchase.

The tangible attributes revolve around the set of design features which make up the actual product. These are set in the context of both an overall style and, for instance, coordination with other garments and accessories to form a "total look". Designers appropriately use fabric, texture, pattern and color, and silhouette for any given season.

The intangible attributes include services intrinsic to the purchase such as after-sales service such as alterations and refund guarantees, and image and reputation of the seller.

The fashion product life cycle

As each new idea passes through the fashion cycle, it goes through a series of stages (Figure 1.2).

- Introduction: new trends are recognized and worn by fashion innovators. At this stage, trends start to appear in designer boutiques and high-end specialty stores;
- Growth: fashion leaders or early adopters pick up on trend and help to popularize. The trend can be found in fast-fashion retailers, contemporary boutiques, and specialty stores;
- Maturity: the trend is interpreted for a mass market lifestyle, and is at the height of popularity. It is widely available to mass-markets at all price points;
- Decline: consumers may still continue to wear the fashion but are no longer interested in purchasing additional items unless it is at greatly reduced prices;
- Obsolescence: the fashion disappears from the marketplace, consumers cannot buy it anymore.

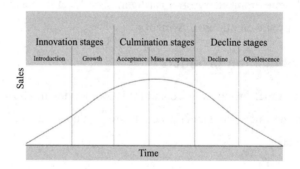

Figure 1.2   Life cycle of fashion

Figure 1.2 shows the bell-shaped curve normally used to depict the product life cycle model. The product life cycle curve is a theoretical model and doesn't necessarily reflect all fashion products. The life cycle may extend over one or several seasons. Being able to recognize and predict the different stages is vital to success in both buying and the selling of fashion goods.

The life cycles of a basic product, a fashion and a fad vary greatly. Basic products change minimally from season to season, and often last for several seasons. **Fashion** is a style that is accepted and used by the majority of a group at any one time. Fashions usually have a slower rise to popularity, reach a plateau with continuing popularity and then decline gradually; often this cycle relates to a season, whether autumn/winter or spring/summer. **Fad**, in contrast, rises meteorically in popularity only to suffer an abrupt decline as they become adopted. Some styles, however, are so enduring that they never go completely out of fashion. These styles are called **classics**. The Chanel-style suit, the five-pocket jean, and the blazer jackets are all examples of classics.

Seasonality

The demand for fashion goods is subject to the vagaries of the weather and the seasons. New ranges are introduced at certain times of the year in the expectation that the weather will be as normal. Thus the industry revolves around a time-based, i.e., seasonal, process in which new fashions are introduced into the marketplace and are adopted by enough consumers to warrant the description of "fashion" in the first place, only to wane eventually in terms of popularity, and become obsolete, i.e., unfashionable. This process is known as planned obsolescence. While some product offerings will remain popular over several or even many seasons, others will fade very quickly. This regenerative process is intrinsic to the fashion industry and is very necessary for its continued survival.

Product lines of apparel are created and styled for each fashion season. A product line is often simply referred to as a "line", which is a group of garments, linked by a common theme like color, print, fabric, or style. Most apparel manufacturers produce between four and six lines every year. For women's wear, they are spring, summer, fall (fall I and fall II), and resort or holidays.

1.2 Fashion industry

1.2.1 Origins of the modern fashion

From ancient Egypt to the beginning of the twentieth century, fashion had been evolving for several thousand years. Fashion had been used by its adopters to show that they were above the common people. Even the inventions of the eighteenth and nineteenth centuries: the spinning jenny, the water frame and the sewing machine, made mass production possible, the modern mass fashion hadn't started until the end of the World War I, in 1918. Mass production methods, coupled with the experience of producing clothing for large armies, led to the possibility of mass markets for clothing. After the World War I, cultural changes, the explosion of the media, and technology developments began to have a great effect on the fashion business. Table 1.2 lists some major developments in the modern fashion industry.

Table 1.2 Some major developments in the modern fashion industry

Time period	Fashion development
1918 onwards	The end of the World War I marked the start of mass fashion
1930s	Film personalities influenced popular clothing
1939–1945	The World War II forced hemlines up because of a shortage of material
1950s–1960s	Freer styles, fewer control garments occurred
1970s–1990s	Fashion business began to grow in many countries, and the media started to become an important influence
1990s	Branded and designer label goods increased
2000 onwards	Electronic shopping began to grow
2002 onwards	Global sourcing increased, fast fashion became popular
2007 onwards	Slow fashion gained more attention

Fast fashion is a contemporary term used by fashion retailers for designs that move from catwalk quickly to capture current fashion trends. Fast fashion clothing collections are based on the most recent fashion trends presented at Fashion Week in both the spring and the autumn of every year.

Slow fashion is a way to "identify sustainable fashion solutions, based on the repositioning of strategies of design, production, consumption, use, and reuse, which are emerging alongside the global fashion system, and are posing a potential challenge to it". It is an alternative to fast fashion in the sense that it promotes a more ethical and sustainable way of living and consuming. It encompasses the whole range of "sustainable" "eco" "green" and "ethical" fashion movement. Some elements of the slow fashion philosophy include: buying vintage clothes, redesigning old clothes, shopping from smaller producers, renting or swapping clothes, making clothes and accessories at home and buying garments that last longer.

1.2.2 The scope of fashion industry

As shown in Figure 1.3, the primary level of the fashion industry is composed of the producers and manufacturers of raw materials. Many raw materials are used in the production of fashion goods. Most of the materials are from the textile industry, including fiber producers, textile mills, converters, and finishers. The second level is to design and manufacture fashion goods. Designers conduct research on fashion trends, interpret them for their audience, and generate salable concepts, which are the creation of a product. An apparel manufacturer buys raw materials and cuts and sews its own garments to make finished goods.

The finished goods are marketed through sales representatives located in fashion market centers and by wholesalers who buy goods for resale to retailers. The final link in the flowchart is the retail distributor. Retailers provide the direct link to the ultimate consumer through stores or direct marketing. In addition to the various producers, manufactures and retailers, the fashion business also includes many auxiliary enterprises that promote fashion, disseminate information, and assist the other segments of the industry.

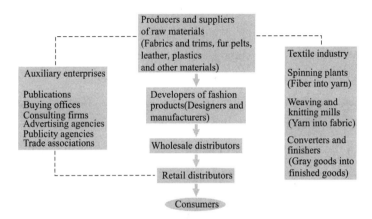

Figure 1.3   Fashion industry flowchart

The scope of the fashion industry is broad and diversified. All segments of the business are interrelated, and all have the same ultimate goal—to identify what goods the consumer will want to buy and to make a profit by producing and selling those goods.

Within the fashion industry, there is enormous variation in the size and structure of businesses serving the needs of customers. From a small business comprising a self-employed knitwear designer to major multinational corporations such as Prada, diversity remains a key feature. With legislative changes and the gradual removal of trade barriers on a global scale and the growth of the Internet, the fashion industry is increasingly a global business.

1.3   Apparel product development and supply chain

1.3.1   Apparel product development

In order for the change which is intrinsic to fashion to take place, the industry must continually create new products. The apparel industry can be classified into two levels: manufacturing and distribution (wholesale and retail). Since the apparel product development become consumer-driven, the process is more conducted by the distribution level (either wholesalers or retailers), who is closer to consumers in the flowchart of the fashion industry.

Product development is a series of steps that includes conceptualization, design, development and marketing of newly created or newly rebranded goods or services. The apparel product development process was once called design, but is now more inclusive. Apparel product developers constantly reevaluate what products they will offer, how those products will be produced, and how they will be marketed and distributed to consumers—hopefully translating change into opportunity and profitability.

The continual process of new product development and resulting change drives the whole industry and answers the demand from consumers for a constant stream of new ideas and offerings. Indeed, it could be argued that without this constant generation and introduction of new ideas into the marketplace, the concept of "fashion" would not exist.

If consumers were not constantly engaged in the process of looking for new products or services to satisfy their emerging needs, the fashion process could not function.

Apparel product development is defined as: the synergistic efforts of trend research, planning, designing, merchandising, and production process to create an apparel product for the intended consumer. Figure 1.4 illustrates the stages of apparel product development. In the following chapters, works to do at each stage will be introduced.

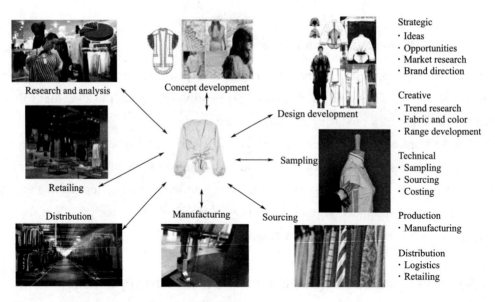

Figure 1.4   Stages of apparel product development

1.3.2   The dynamics of product development

Apparel product development processes vary, depending on:

- **Who** is doing the developing—wholesalers or retailers;
- **What** kind of products they are developing—basics, seasonal, or fashion forward; whether they are products that require high skill or high technology as opposed to those that are low skill, low technology;
- **Where** the products will be distributed—regionally, nationally, or globally;

• **How** the products will be sold—in stores, television, or online;

• **When** the products need to be delivered.

All these elements of the marketing mix are dynamic and nonlinear, and create a chaotic process. The field of apparel is a captivating industry. It is one in which superstardom (Internet celebrity) success may be achieved overnight, take years to accomplish, or never happen. The apparel industry is intensively competitive, and its fast pace and continual product change is thrilling to many people.

To maximize efficiency and maintain a competitive edge, product developers must identify their **core competencies**—the things they do best—and partner with specialists to develop and distribute the goods offered under their brand name. For example, the core competencies of businesses in developed countries tend to focus on product development processes that add the most value— trend research, design, marketing, and distribution; they outsource apparel manufacturing to partners in developing countries where labor is much cheaper. The word sourcing means finding the best production available at the cheapest price. As fabric and garment manufacturing moved away from the developed countries of the West, to the developing nations of the East, sourcing in apparel product development become global.

It is not unusual to be able to buy the same garment in Japan, London and New York that was:

• made up in Turkey;

• with sewing threads from China;

• from fabric woven in Korea;

• from yarns manufactured in Italy and Greece;

• from fibers made in Germany.

1.3.3 Apparel supply chain

Nowadays, apparel product development is a defining step in the complex network of suppliers and service providers involved directly or indirectly in fulfilling customer demand for apparel; this network is referred as the apparel supply chain. In short, a supply chain is a set of firms that make and distribute goods and services to consumers. An apparel supply chain consists of fiber, textile, and findings (trim, threads, label) suppliers; apparel products developers; manufacturers and contractors, and some channels of apparel distribution. This network also includes auxiliary business, such as design bureaus, software providers, sourcing agents, factors (credits agents), patternmaking services, testing labs, consultants, warehouses, shipping companies, and advertising agencies. Figure 1.5 shows a conventional intimate apparel (IA) supply chain.

Sourcing is the continuous review of the need for goods and services against the purchasing opportunities that meet quantity, quality, price, sustainability, and delivery parameters, in order to leverage purchasing power for the best value. Any goods that a brand can't produce or functions that

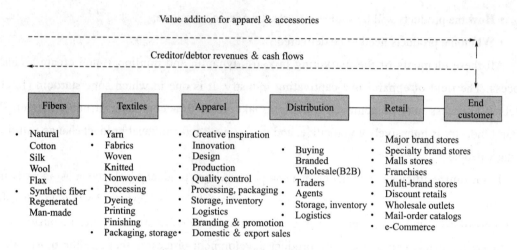

Figure 1.5 Conventional intimate apparel (IA) supply chain

a brand can' t perform cost-effectively are sourced out to other vendors. With proper oversight and communication, the expertise of each supply chain participant or sourcing partner contributes to the efficiency of the entire chain and increases the value of the resulting products.

For each product developed, a unique product supply chain is assembled from a network of suppliers or vendors that can meet the requirements for the order. Individual product supply chains are determined by the specifications for that product (i.e., styling, fabric, construction, and quality requirements), as well as considerations such as the size of the order and the required delivery date. Each product supply chain lasts only as long as that product is produced, but the individual links of the chain may become part of numerous supply chain variations within a given design season, depending on how many diverse styles they have the core competencies to produce.

Supply chain structures

Vertical integration is a strategy that seeks to consolidate a supply chain by acquiring a company at another stage in the supply chain. When a firm engages in service integration, they are performing multiple activities within the supply chain channel. Figure 1.5 shows a typical Ⅵ model of supply chain in the fashion industry, consisting of design, fabric/trim production, garment manufacture, and distribution. The advantage of this type of supply chain is that it allows apparel companies to have the intrinsic ability to control all aspects of their process. However, retailers and brands increasingly move toward a design/source/distribute (DSD) model by focusing on their core competences of design, branding and retailing, with the production function outsourced to global networks of independent suppliers.

Horizontal integration is an alternative strategy. Horizontal integration prioritizes the acquisition of companies that make or sell similar products in order to expand market penetration and reduce competition; it may be used to acquire brands at the same price point or to penetrate multiple points.

Regardless of whether the acquisition strategy focuses on vertical integration or horizontal integration-collaboration is key. A collaborative supply chain is an interactive network of manufacturing specialists who join forces operationally to integrate complementary resources to respond to a market opportunity through the creation of a particular product. Today's digital technologies

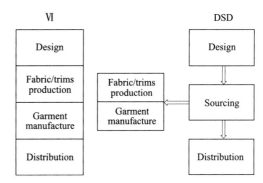

Figure 1.6   Supply chain models in the fashion industry

make it possible for companies, regardless of geographic location, to coordinate their core competencies into a single effort to achieve competitive advantage (Figure 1.6).

Supply chain management

Supply chain management is the timely integration of information and processes among suppliers, manufactures, warehouses, and stores to ensure a competitive advantage. It refers to the management philosophy that integrates all of the business functions within a flexible supply chain. Supply chain management software helps companies coordinate the flow of materials, information, and finances as the product moves through the chain.

◎ Summary

Fashion is about change and creation. Fashion products are diverse and seasonal. Fashion business includes all of the industries necessary to produce and market fashion goods to the consumer. Apparel product development is not just about design, it is the synergistic efforts of trend research, planning, designing, merchandising, and production process to create an apparel product for the intended consumer. Product development must respond to rapidly evolving consumer preferences by leveraging a complex global supply chain.

◎ Key terms

Apparel supply chain 服装供应链

Auxiliary enterprise 辅助行业

Basic product 基本产品（基本款）

Bespoke 订制

Core competency 核心竞争力

Fabric 面料

Fad 狂热

Fashion 时尚、时装

Fast fashion 快时尚

Fashion forward product 时尚前瞻商品

Fiber 纤维

Finishing 后整理

Global sourcing 全球采购

Horizontal integration 横向整合

Knitting 针织

Mass fashion 大众时尚

Ready-to-wear (RTW) 成衣

Retailer 零售商

Seasonality 季节性

Slow fashion 慢时尚

Sourcing 采购

Supply chain 供应链

Trim 辅料

Vertical integration 纵向整合

Weaving 机织

Wholesale 批发

Yarn 纱线

◎ Product development team members

The following product development team members were introduced in this chapter:

Apparel associates（服装相关者）: people who work in or support product development. These individuals consist of company employees, suppliers, consultants and service providers.

Designer（设计师）: a creative individual who generates salable concepts.

◎ Activities

Visit a shopping mall, choose a global brand store. Look at the country of origin on label from a variety of items of the brand. How many different countries of origin can you identify? Discuss how the country of origin might have an impact on the product development process for this brand relative to cost, lead time, and quality.

◎ Review or discussion questions

1. What are the core competencies of companies?

2. Identify some types of slow fashion and discuss how they affect our life or environment.

◎ References

[1] KEISER S, VANDERMAR D, GARNER M B. Beyond Design: The Synergy of Apparel Product Development[M]. 4th Edition. New York: Fairchild Books, 2017.

[2] 张玲 . 图解服装概论 [M]. 北京：中国纺织出版社 . 2005.

[3] EASEY M. Fashion Marketing[M]. New Jersey: Wiley-Blackwell, John Wiley & Sons, Ltd., 2009.

◎ Useful websites

[1] Supply chain brain, www.glscs.com.

[2] New York Times, www.nytimes.com/business/smallusiness/.

[3] WWD, www.wwd.com.

[4] WGSN, www.wgsn.com.

[5] Apparel Search Fashion Directory, apparelsearch.com.

[6] Target Corporation, www.target.com.

[7] Li & Fung, https://www.lifung.com/.

◎ Appendix

Table A1.1   Classifications of women's wear

Classification	Descriptors
Separates and coordinates	Separates are designed to mix and match. Pants, tops, jackets, blouses. Coordinates are designed to be merchandised as outfits or multiples
Knitwear	Because of the requirements of fabrication, knit products form their own category. Sweaters, cardigans
Active sportswear	Evolved into separate category with many specializations dedicated to specific sports. Most are separates: pants, tops, jackets
Outwear	Coats, suits, rainwear, jackets
Dresses	One or two-piece styles and ensembles with a coat or jacket
Eveningwear	Bridal and bridesmaids attire, dressy apparel, cocktail dresses and formal attire
Intimate apparel	Lingerie, foundations, loungewear and robes
Uniforms and aprons	Housedresses, job and career apparel other than office attire often defined by job function
Swimwear and beachwear	Swimsuits, bikinis, beach kaftan, beach dress
Maternity	Pregnancy clothes: dresses, leggings, bras, tops, jackets, coats, pants, skirts, pajamas, swimsuits

Table A1.2   Classifications of men's wear

Classification	Descriptors
Tailored clothing	Tailored suits, overcoats, topcoats, sport coats, slacks
Casual wear (casual Friday)	Evolved from dressing down in business. Casual suits, contrasting coats and trousers, sweaters, knit shirts, pants

continued

Classification	Descriptors
Men's furnishings	Shirts, ties, underwear, socks, sleepwear, robes, scarves, gloves, hats
Active sportswear	Growing category with many specialized products dedicated to specific sports
Outerwear	Casual jackets, windbreakers, snowsuits, ski jackets, parkas
Work clothes	Uniforms often for specific occupations, work shirts and pants, overalls, jeans
Rainwear and coats	Tailored outerwear, raincoats

Table A1.3　Children's wear categories

Classification	Descriptors
Infant wear	Layettes including: shirts, onesies, gowns; Stretch coveralls, diaper pants; Outerwear, knit sweaters, booties, hats
Toddler wear	Boy and girl sets; Shorts, pants, skirts, overalls, shirts
Sportswear	Pants, shorts, skirts, overalls; Blouses, shirts, jackets, sweaters; Swimwear
Dresses	Tailored and party dresses; Jumper and blouse sets
Outerwear	Coats, jackets, parkas, snowsuits
Lingerie and sleepwear	Robes, pajamas, nightgowns; Slips, undershirts, panties, hosiery

Chapter 2

Planning for Success

课题名称：Planning for Success

课题内容：成功之路

课题时间：4课时

教学目的：

1. 理解策略性计划在产品开发过程中的重要性。

2. 能辨识公司的使命和愿景。

3. 能理解市场定位和品牌分类。

4. 理解市场和营销是服装企业的主要功能。

5. 理解时间与行动日历驱动着服装产品的开发过程。

2.1　Business planning

A **business** is an enterprise in commercial, industrial, or professional activities to achieve goals. Whether launching a new business or operating an established firm, planning is integral to product development.

For a business organization to work efficiently and profitably, it must identify the jobs or tasks that need to be done. These tasks are grouped into the firm's **basic business functions**. Though the functional requirements of every business vary, the functions of textile and apparel firms generally include (but not limited to) marketing and sales, merchandising, supply chain management, information technology, and finance.

Central to the decision-making process of any apparel firm is its overall **business plan**—its aim, objectives, and strategy for the future. Emphasis on the components within the plan depends heavily on whether the basic core business of the firm is manufacturing or retailing and whether the business will be marketing itself as a wholesaler, a retailer, or a combination of the two. The intent of the business plan is to provide structure to an ongoing process that is constantly being adjusted to respond to consumer demands. The highest level of planning in a business is strategic planning.

2.2　Develop company strategy

Business plans are made up of different elements or components. At the top level of planning is strategic planning. The big picture is defined and business goals are established by the firm's executive team. To attain success, company associates must be convinced that the products will sell, make a difference, and be superior to existing products. By having a strategy, a company can efficiently serve a market, anticipate market behavior, and have the infrastructure necessary to meet market demands. A firm's missions, vision, and values are the compass of strategic planning.

2.2.1　Missions, vision and values

A **mission** statement is part of the DNA of a successful company, and it reflects the company's heritage by articulating why it is in business, the customer it serves, and what makes it special. It is a written statement that defines why a company exists (its purpose), what it wants to accomplish (its function), and how it will implement its vision.

The goal of a mission statement is to help employees determine a company's purpose and align employees in the same direction. The mission statement becomes part of the company's corporate story that can be conveyed to all stakeholders-employees, investors, those in the communities in which it does business, and supply chain partner, and helps to align them in the same direction.

A **vision** statement outlines the organization's goals for the future and how it wants the world to see its brand. Executives plan their company's future direction and goals, and write in a vision statement. Embedded in a vision statement are the executives' knowledge and foresight, which are used to project the company's philosophy, beliefs, and brands. The objective of a vision statement is to provide a company's employees with a direction and clear goals. A vision must enable employees to feel a sense of pride, accomplishment, and value about what they contribute to the organization.

Example: Christian Dior

Vision: Christian Dior has continued to assert its vision through elegant, structured, and infinitely feminine collections. Dior had just one obsession: to allow women to rediscover joy, elegance and beauty.

Mission: Dior revolutionized the conventions of elegance and femininity, designing collections infused with dreams and enchantment. It uses flowers, colors and art all in their own way, to build a legacy that the House of Dior upholds and to preserve this timeless and inimitable attitude.

A vision is a future projection and a mission is what a company implements to attain its vision. Some companies just combine their mission and vision statements. A company's mission and vision statements reflect its **core value**—what it holds to be important as a business. Whether these values are spelled out or merely implied in the statements, they must be authentic. Mission, vision and values form the foundation of a company's strategic plan.

2.2.2   Market positioning strategy

Positioning and perceptual mapping

A fundamental part of serving a particular target market is product positioning. It involves adapting a marketing mix so that the target market will see the product as meeting their wants and needs. To position a product or a brand, it is necessary to know how consumers perceive it in that category. The main method of determining a market position is the use of marketing research to construct a perceptual map of the market. A perceptual map is the consumers' view of the market, where consumers provide the main dimensions or criteria for making judgements.

In positioning a product, it is necessary to consider the relationship of the product's position to that of competing products. A positioning map is a two- or three-dimensional diagram of how consumers perceive various brands in terms of specific attributes. In fashion businesses, positioning refers to how the garment relates to others in the marketplace in terms of style, complexity of design, fabric, quality, and price. For example, if a brand defines itself as fashion forward, then it must include garments that show fashion leadership and fewer basics. Brands position their products by identifying elements such as functional use, quality level, size range, rate and type of product change, product characteristics and price.

An important point to be made about the perceptual map is that it is hypothetical and may have little to do with the realities of style and pricing. The essence of positioning to succeed is to get the product right in terms of customer needs and expectations and then to tailor the image of the firm's marketing offering to meet the aspirations of the chosen market segment.

Product positioning by price point

In the fashion industry, products are segmented by price classifications. Before designing apparel and textile products, a firm must determine the price point at which it wants to compete. **Price points** are prices at which demand for a given product are supposed to stay relatively high. The target price point is determined by understanding the price range that the intended target customer is willing to pay. Target price categories help to identify competitors and appropriately position the product in the marketplace. Table 2.1 lists price point categories of women's wear.

Table 2.1 Price point categories of women's wear, from most to least expensive

Price point	Description	Examples
Haute couture	Made-to-order garments. They are made of the highest quality fabrics and probably by some handwork. The consumer goes to the designer's salon to be fitted for the garment	Chanel Couture, Dior Couture, Jean Paul Gaultier
Designer	The most expensive garments that can be bought off the rack are known as designer ready-to-wear(RTW). They are beautifully designed, impeccably made, and use high-quality fabrics	Prada, Yves St. Laurent, Donna Karan, WOOYOUNGMI, Armani, Gucci, Xander Zhou, XU ZHI
Bridge	A price point slightly lower than designer. It consists of designer diffusion collections—strong aesthetics but in lower quality, thus lower prices	Dana Buchman, Anne Klein, Dolce & Gabbana, Michael Kors, Coach
Contemporary designer	This category includes many new designers who target a younger, fashion-savvy customer	Tracy Reese, Marc by Marc Jacobs, MA SHA MA, UMA WANG
Better	Products with wide market appeal, often the highest price point available in department stores	Calvin Klein, Lauren, Banana Republic
Moderate	Affordable fashion products for the large and price-conscious market. Styling appeals to more mature customers	Chaus, Sag Harbor, Norton McNaughton, J.H. Collectibles
Low-end contemporary	A relatively new category that offers fast fashion at a low price point. They are trend-driven products of low quality and sold in high volume at low price points	BCBG

2.2.3　Branding

A **brand** is a name, term, design, symbol or any other feature that identifies one seller's goods or service as distinct from those of other sellers. Product developers and retailers use brands as a tool to communicate with customers and drive repeat business.

Branding is a competitive strategy a company uses to identify and communicate its products or services and provide customers with assurances of a level of quality and consistency of standard. It is the process of attaching a name, image, or reputation to a product, idea, or service. Brand names, logos, symbols and slogans are generally proprietary and legally protected through intellectual property laws.

To plan for brand development there need to be an overarching focal point. Much like the mission and vision statements, many companies create brand statements. A brand statement is a touchstone to communicate the intrinsic value of the brand and provides direction for where the brand is heading in the market. Here are questions to examine when developing a brand statement:

What is the brand message?

How does the consumer experience the brand?

What brand concepts are compelling and differentiating?

Brand image is defined by the consumer's set of assumptions and feelings about products and/or services provided under the brand name. Effective branding should be organized around a coherent purchase and repurchase of products or services from the same company. This can be achieved through packaging, advertising, store environment; promotion, customer service, word of mouth; and also through association with a lifestyle, celebrity, popular cause, or event. Figure 2.1 shows a hypothetical perceptual map of some fashion brands.

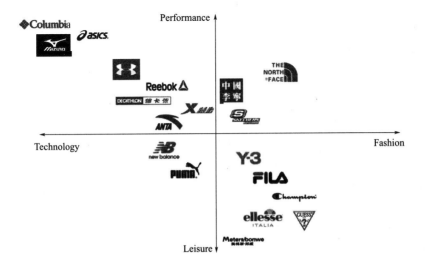

Figure 2.1　A hypothetical perceptual map of some fashion brands

2.2.4 Competitive advantage

The apparel business is classified as a mature industry. Mature industries produce products that are characterized by relatively stable sales from year to year and by a high level of competition. Given a relatively stable population rate, the domestic apparel market is saturated with suppliers who must compete by maintaining a competitive advantage. A company must define the brand's competitive edge.

Competitive edge/advantage refers to attributes that allow an organization to outperform its competitors. These attributes make a product or brand unique or distinct from others. For fashion products or brands, the competitive edge can be lower price, unique product features or aesthetic, unique fit, high-tech, or speed-to-market.

2.3 An overview of fashion business

2.3.1 Marketing and merchandising

The main functions of a fashion business are marketing and merchandising. **Marketing** includes all of the activities involved in conceiving a product and directing the flow of goods from producer to the ultimate consumer. Activities of marketing include product development, pricing, promotion, and distribution. Successful marketing is customer driven: the firm must have a product that consumers perceive as desirable, and the product must be presented to potential customers in a way that makes them want to buy it. For a firm to do this, the organization must be marketing-oriented at all levels of the business.

A company plans its entire operation around satisfying the consumer. The first step is to define a company's target customers. Next, the company identifies these customers' needs and wants. Products that will meet the needs and desires of target customers are then developed or selected.

Merchandising is like a functional umbrella where most product development activities occur. The **vice president** of product development, design directors, merchandisers lead product development associates in planning and developing design and business goals for an identified target market. A **design director** guides the design team in developing creative concepts. The primary role of a **merchandiser** is business planning, involves developing and analyzing. In the following text, the merchandising planning is outlined with a focus on the merchandiser's role in apparel product development.

2.3.2 An overview of a merchandise planning

Market research

The merchandising team visits retail stores, communicates with buyers, and shops the competition.

They monitor sales reports and the brand's social media in order to stay on the top of customer responses. They work with the design team to interpret trends in color, fabrication, trim and silhouettes for their target customer to ensure that their product line has a competitive edge. In a word, they are the eyes of the brand who link the product line to customer needs.

The line plan

It is the task for merchandisers to develop a business plan that identifies the resources needed to meet profit, sales, and margin objectives for a specific season, by company division. This plan begins at "top level" to estimate how much merchandise a company plans to sell ahead of the season. To develop this plan, merchandisers analyze last year's sales; plan sales increases/anticipated reductions; targe first price, volume, and desired margin. Upon this analysis, a line plan, also known as a range plan, is drafted and sales and margin goals are set for categories. Some companies break down the budget by categories.

A line plan is an overview of the collection with all of the design and financial parameters set out. The plan outlines the specifics of the collection from how many styles you will have (the breadth of the product mix), to what fabrics and colorways will be used (the depth of the product mix). This line plan will direct the design team so that they can put their energies into designing items that are necessary in the line.

Time and action calendar

Working from the line plan, the merchandising team develops a time and action calendar, a robust tool for aligning and coordinating design, technical development, and specification processes. The time and action calendar establishes and assigns responsibilities for starting dates and deadlines for completion of all key events in developing a seasonal line. It starts with fixed dates for line release and trade shows, as well as delivery dates required to prepare for special promotions. By working backwards from these dates, the merchandiser tracks major milestones, including market research, design development, sourcing, first samples, fittings, production samples, wholesale orders, production, and delivery. Figure 2.2 illustrates three variations of the product development calendar.

Assortment planning

Assortment planning further refines the line plan, in terms of variety, volume, diversity, and distribution. The objective of assortment planning is to select and plan products to maximize sales and profit for a specified period of time.

A cohesive line is bound into a group through a centralized theme to make a single statement. Using the line plan as a template and concept boards as a creative guide, designers begin to develop silhouettes. An assortment plan shows how many styles, fabrics, colors and sizes that will be

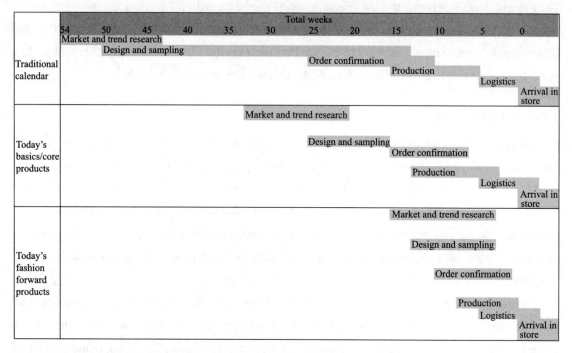

Figure 2.2 Comparison of three product development calendars

included in the line to meet customer demand.

Line review

Once the line has been developed, it is presented to top-level executives; sales representatives (wholesale brands) or buyers (private brands), and sometimes, key customers for review. This evaluation ensures that styles with little sales apparel aren't further developed.

Line development

After the line review, all approved styles are produced to create a sample or prototype of the design. Upon fit approval of the sample, the style becomes ready for production. Final costing is executed. The technical package, which specifies the colors, fabrics, notions and trims, garments construction, measurements, and material and garments testing required for each style is developed.

Pricing

The price levels for a product line can, along with other elements of the marketing mix, decide the success or otherwise in attracting certain target markets. By using a method of either cost-plus pricing, or market-based pricing, or combination of both, merchandisers price garments in the line to meet gross margin targets.

Production contracts are awarded to companies that can provide acceptable quality and reliable delivery at a fair price. If a garment is determined too costly to make, its design may be modified or it may be dropped from the line. Once all production orders have been placed, merchandisers work with the sales team to set wholesale and/or retail prices.

Line presentation

A line presented as a document with detailed production description, called a **line sheet**.

The products shown in **line presentation,** are fully defined and priced. A finalized line described by a technical package is presented to sales representatives or buyers to quantify orders. Once the line is quantified, sourcing takes over the management of production.

Production control

Mass production is the actual development of product in bulk quantity. Based on the cumulative wholesale orders, an apparel company will contract to purchase fabric and components. Most often a cut-and-sew contractor or a cut make and trim (CMT), a manufacturer who cut the fabric, makes and trims the garment, is used to facilitate mass production. Once all of the sourced fabrics, materials, and trims are secured, production is scheduled. Production control oversees approvals and keeps track of progress to ensure that goods will be delivered on time.

◎ Summary

The success of a business depends on developing and implementing a strategic plan that provides focus for all activities of the firm. The strategic plan is based on the mission and vision of the firm. Branding is a competitive strategy a fashion company often uses to identify and communicate its products. Merchandising planning is the essential function for a company developing fashion products. It is the line plan that links strategic planning objectives to creative, technical, production planning, as well as to sales and marketing goals.

◎ Key terms

Assortment plan 产品组合计划

Basic business functions 企业机能

Brand image 品牌形象

Branding 品牌化

Business plan 商业计划

Buyer 买手

Competitive edge/advantage 竞争优势

Core value 核心价值

Cut-and-sew contractor 裁剪与缝纫承包

Cut make and trim (CMT) 裁剪与缝纫外包

Haute couture 高级时装

Line plan/range plan 产品系列计划

Line sheet 产品目录

Marketing 市场营销

Merchandiser 营销策划师

Merchandising 营销策划

Mission statement 使命宣言   Range plan 产品系列规划
Perception 定位   Sales representatives 销售代理
Positioning 认知   Time and action calendar 时间与行动日历
Price point 价位层次   Vision statement 愿景宣言

◎ Product development team members

The following product development team members were introduced in this chapter:

Entrepreneur/President（总裁）: a top executive manager. His or her responsibilities include overseeing drafting of and adherence to vision and mission statements, and policy; guiding the company's direction and growth; and addressing critical issues facing product lines.

Merchandisers（营销策划者）: individuals who combine creativity with business planning. They are responsible for merchandise planning. Their responsibilities also include analyzing profitable and unprofitable designs and ensuring that the product development associates adhere to the time and action calendar.

Vice president of product development（产品开发副总裁）: an individual at the top of the product development organization. His or her responsibilities include writing strategy, overseeing continuity of product across product lines, analyzing product line profitability and critical product line issues.

◎ Activities

1. Choose several of your favorite brands. Go to their websites and look up their vision, mission and values (VMV). Analyze whether the company's VMV are clear and well defined. Look at the company's products. Do they reflect the company's VMV? Why or why not?

2. Look at one sportswear brand's advertising and its products for one product line, identify several consumer purchase attributes.

◎ Review or discussion questions

List some of your favorite fashion brands, identify what differentiates them from their competition in the marketplace.

◎ Semester project

Objective: In small groups, make preproduction business decisions for a fictitious apparel company, and develop a business plan for this company.

Semester project I : Business strategy report—Part 1

1. Create a fictitious apparel company by determining the following:

Products: product categories (men's, women's, or children's), price point categories.

Brand: name, symbol, vision statement, mission statement.

2. T&A: Develop a timeline for one seasonal line for this business, beginning with the delivery of the product line to stores and working backward to establish the time allotment for each of the functions identified in this chapter.

◎ References

[1] KAPFERER J N. The New Strategic Brand Management [M]. 3rd Edition. London: Kogan Page, 2007.

[2] KEISER S, VANDERMAR D, GARNER M B. Beyond Design: The Synergy of Apparel Product Development [M]. 4th Edition. New York: Fairchild Books, 2017.

[3] CARR M G, NEWELL L H. Fashion Entrepreneurship, the Plan, the Product, the Process [M]. London: Bloomsbury Publishing Inc, 2014.

◎ Useful websites

[1] Apparel News-Trade Show Calendar, http://apparelnews.net/events.

[2] Apparel Search Fashion Directory, apparelsearch.com.

[3] Marketing scoop, www.marketingscoop.com.

[4] NASDAQ, www.nasdaq.com/reference/BarChartSectors.stm?page=sectors&sec=textile-apparel~clothing.sec&level=2&title=Textile%2dApparel+Clothing.

[5] Start a fashion business, www.startafashionbusiness.co.uk/quiz-can-you-run-fashion-business.html.

[6] The balance small business,www.thebalancesmb.com.

[7] Entrepreneur, www.entrepreneur.com.

[8] Apparel Search Fashion Guide, https://www.apparelsearch.com/terms/m/market_week_term.html.

◎ Appendix

Table A2.1   Men's wear release

Month	Major show(sponsor)	Location	Website	Product type
Jan.	Milano Moda Uomo (Camera Nazionale della Moda Italiana)	Italy	www.cameramoda.it/it	Men's luxury ready-to-wear and designer clothing
Jan.	Moda à Paris (Fédération Française de la Couture)	France	www.modeaparis.com	Designer ready-to-wear collections
June	Milano Moda Uomo (Camera Nazionale della Moda Italiana)	Italy	www.cameramoda.it/it/	Men's luxury ready-to-wear and designer clothing

Table A2.2   Women's wear release

Month	Major show(sponsor)	Location	Website	Product type
Feb.	Prêt à Porter Paris	Italy	www.cameramoda.it/it	Moderate, better, and contemporary clothing
	London Fashion Week	England	www.londonfashionweek.co.uk/?display=flash	Designer ready-to-wear collections
	Milano Moda Uomo (Camera Nazionale della Moda Italiana)	Italy	www.cameramoda.it/it/	Women's luxury ready-to-wear and designer clothing
Feb.–Mar.	Moda à Paris (Fédération Française de la Couture)	France	www.modeaparis.com	Designer ready-to-wear collections
Mar.	Mercedes Benz Los Angeles Fashion Week	United States	www.mbfashionweek.com/losangeles/	Los Angeles designer apparel, sportswear
Sept.	London Fashion Week	England	www.londonfashionweek.co.uk/?display=flash	Moderate, better, and contemporary clothing
	Milano Moda Uomo (Camera Nazionale della Moda Italiana)	Italy	www.cameramoda.it/it/	Women's luxury ready-to-wear and designer clothing
	Prêt à Porter Paris	France	whosnext.com	Moderate, better, and contemporary clothing
Sept.–Oct.	Moda à Paris (Fédération Française de la Couture)	France	www.modeaparis.com	Designer ready-to-wear collections

Table A2.3   Children's wear release

Month	Major show(sponsor)	Location	Website	Product type
Jan.	Pitti Immagine Bimbo	Italy	www.pittimmagine.com/corporate/fairs/bimbo/html	Children's wear
Jan.	Children's Club (ENK International Trade Events)	United States	www.coteriefashionevents.com/en/home.html	Children's wear
June–July	Pitti Immagine Bimbo	Italy	www.pittimmagine.com/corporate/fairs/bimbo/html	Children's wear
	Children's Club (ENK International Trade Events)	United States	www.coteriefashionevents.com/en/home.html	Children's wear

Chapter 3

Consumer Markets

课题名称：Consumer Markets

课题内容：消费市场

课题时间：4课时

教学目的：

    1. 了解消费行为及其影响因素。

    2. 理解市场调研的重要性及其方法。

    3. 掌握市场细分的主要方法。

    4. 学习锁定目标市场。

3.1 Consumer behavior

3.1.1 Fashion consumer decision

Fashion firms depend upon customers making repeat purchases and the key to such loyalty is the satisfaction of customers' needs with garments which are stylish, durable, easy to care for, comfortable, perceived value for money and all the other criteria deemed relevant by the buyer. For this reason, fashion associates should readily appreciate the need to understand the customer's perspective and behavior.

A **consumer** is one that buys goods for consumption and not for resale or commercial purpose. One main way of examining consumer behavior is to take the view of the consumer as a problem solver. The requirement for clothing is seen as a problem for the consumer to solve. The problem-solving perspective raises many questions that will be addressed; they concern the types of decision fashion consumers must make, the various stages of the decision process consumers progress through and major factors that influence those decisions.

The purchase of a garment seems to be as just one decision, to buy or not to buy. However, this decision can be broken down into several separate decisions that collectively comprise the buy or no buy decision. For instance, consumers must make a series of smaller decisions on the following matters:

- How to find out about new styles?
- When to buy?
- Where to buy from?
- Whether to shop alone or accompanied?
- What style, color and size to buy?
- Whether to try the garment on?
- Whether to order an out of stock size or color option?
- How many items to buy?
- Will any accessories need to be purchased?
- How to pay?
- What to do if the product is unsatisfactory?
- What will be the reaction of significant others to the purchase?
- Whether or not to purchase online?
- If buying online, how to arrange delivery?

3.1.2 Fashion consumer decision–making process

Figure 3.1 illustrates the consumer decision process and the main explanatory variables related to the decision process.

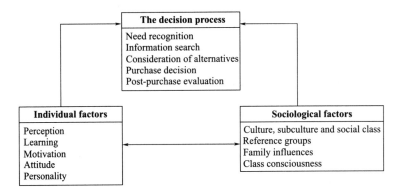

Figure 3.1   A model of consumer behavior

Need recognition

First, recognition occurs when a consumer becomes aware that a need for clothing arises. This may be triggered by garments wearing out, comments from others about how unfashionable existing garments are, a change of social status prompting or facilitating purchase or a change in aspirations or taste.

Information search

Having become aware of a need, the consumer reflects on the situation and can decide to proceed with the purchase process and collect information, defer the purchase or conclude that the problem is insignificant or cannot be solved. In deciding to proceed, the consumer reviews information already held in memory, and/or acquires external information sources to be consulted.

Consideration of alternatives

When sufficient information is held by the consumer about possible solutions then evaluations take place and a choice is made. The process of making an evaluation may involve the mental ranking or rating of alternatives or simply eliminating items that fail to meet a certain threshold. The nature of the evaluations varies from individual to individual. Some consumers have extensive repertoires of buying criteria, whereas others have limited and often vague mechanisms for making decisions.

Purchase decision

After evaluating the possible alternatives, the consumer makes a purchase decision. This step includes deciding whether to buy and if so, what to buy, where to buy, and when to buy. The consumer's choice can depend in part on the reason for the purchase. For instance, when some guy wants to buy some T-shirts to wear around the house, he is heavily influenced by price; but when he is on vacation, he likes to pick up a T-shirt with a clear design as a souvenir, even though it costs twice as much. The consumer's mood can also have a big influence on the product choice.

Post–purchase evaluation

After a purchase, consumers formally or informally evaluate the outcome of the purchase. In particular, they consider whether they are satisfied with the good they bought and the experience of making the purchase. A consumer who repeatedly has favorable experiences with the same purchase decision may develop loyalty to the brand purchased.

3.1.3 Individual factors affecting consumer behavior

The above discussion of the decision process has shown the need to consider factors beyond the immediate concerns of the consumer. The factors to be considered may be grouped under the broad headings of individual and sociological factors. Psychological factors are taken from the study of individual behavior while sociological factors are based on the understanding that much consumer behavior takes place as part of a group process and involves social interaction and patterns of influence.

The consumer's decision is influenced by his or her perception, motivation, learning, attitude, and personality.

Perception

Perception refers to the way people gather and record information. In addition to exposure to sensory stimuli, the perception process includes attention, interpretation and retention.

Learning

The fashion consumer is not born with a knowledge of fashion brands, of criteria for judging garments, a knowledge of stores or prices, preferences for certain styles or fabrics or even how to care for garments. All this information has to be learned. Consumer learning is any relatively permanent change in buying behavior that is a result of practice or experience.

Motivation

When consumers perceive that they have a need, the inner drive that propels them to fulfill the need is called motivation. A widely cited classification system for motivation was developed by a psychologist, Abraham Maslow. Figure 3.2 shows Maslow's five-

Figure 3.2   Maslow's hierarchy of human needs

level hierarchy of human needs. He postulated that what motivates people to act is unfulfilled needs and that people meet certain basic needs before being highly motivated to meet other needs.

Attitudes

When people are motivated to satisfy a need, the way they meet that need depends on their attitudes toward the various alternatives. An attitude is the combination of a person's belief about and evaluations of something, leading to a tendency to act in a particular way. It consists of three components: cognitive, affective, and behavioral. Attitudes are formed after some thought, they are learned, and they occur within given circumstances. Figure 3.3 shows a simple model of the link between attitudes and buying behavior.

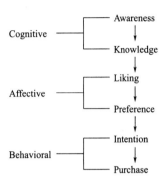

Figure 3.3   The link between attitudes and buying behavior

Personality

Personality refers to individual differences in characteristic patterns of thinking, feeling and behaving. Those individual psychological characteristics that routinely influence the way people react to their surroundings make us unique from other people. Characteristics, such as whether we are more extroverts or introverts, passive or active, leaders or followers, are part of our personality. Two main approaches to personality are used for fashion consumers: psychographics and self-concept.

Psychographics. **Lifestyle** is the manner in which people conduct their lives, including their activities, interests and opinions (AIOs). The process of identifying various categories of lifestyles is called psychographics. Psychographics usually focuses on creating profiles of consumers within each category of lifestyle. Typically consumers are asked a large number of questions, usually based on a Likert scale, of general and specific questions relating to AIOs. Examples of such questions are "Do you like shopping online?" (general) and "How often do you shop online?" (specific). The answers to these questions are then analyzed together with demographic, purchasing and media data about consumers to derive distinct groupings, or types who have AIOs in common.

Self-concept theory addresses personality by looking at the self, our characteristics and attributes, that is self-concept. The dimensions of self-concept include:

- The self-image, who we think we are;
- The ideal self-image, who we would like to be;
- The social self-image, how others see us;
- The ideal social self-image, how we would like others to see us.

Clothing is an obvious way in which a consumer may express him- or herself and show how they would like others to judge them. Self-images are affected by many factors, but most notably by age and social class. Perceived opportunities and/or the lack of opportunities can influence not only images of the self, but also the value placed upon the self, namely self-esteem.

3.1.4 Sociological aspects affecting consumer behavior

Consumers are social creatures, who form groups and interact in relation to goals. Consumer behavior when viewed from a sociological perspective is more than a simple aggregation of individual acts, for the patterns and processes of both individual acts and wider social changes are profound in their impact on fashion marketing. The following social dimensions of consumer behavior relate particularly to the process of influence over choice and to the basis for segmenting markets.

Culture, subculture and social class

Each of us is a member of a culture, various subcultures, and a social class. Those groups influence our buying behavior by providing direct messages, and also indirectly by helping to shape the values and attitudes that influence purchase decisions.

A **culture** is the complex of learned values and behaviors that are shared by a society and are designed to increase the probability of the society's survival. A **subculture** is a segment within a culture that shares distinguishing values and patterns of behavior that differ from those of the overall culture. **Social class** refers to people who have not only similar income levels, but also comparable wealth, skill and power. The most widely used system in the fashion market is the National Readership Survey using A, B, C1, C2, D and E. Table 3.1 describes these classes.

Table 3.1 Classification of social classes

Class	Occupation
A: Upper middle class	Higher managerial, administrative and professional
B: Middle class	Intermediate managerial, administrative and professional
C1: Lower middle class	Supervisors, clerical or junior managerial
C2: Skilled working class	Skilled manual workers
D: Working class	Semi/unskilled manual workers
E: Subsistence level	Includes most pensioners and the unemployed

Reference groups

Besides sharing the values of a culture, subculture, social class, consumers consider or consult with various groups when making purchase decisions. A group may be defined as two or more people who bear a psychological relationship to one another and who interact in relation to a common purpose.

Consumers use various groups as a reference point for evaluating their own beliefs and attitudes.

Family influences

A family is defined as a group of individuals who line together and are related either by blood, marriage or adoption. The family is the basic social group and the main mechanism by which social values and aspirations are transmitted. Thus, the family is among the most important influence reference groups for most consumers. Paying attention to the role of the family is important because family members often make purchase decisions and purchase for one another.

Class consciousness

Class consciousness reflects individuals' desire for social status. It is seen in their choice of friends, where they choose to live, and how they spend their disposable income. In the past, the purchase of luxury items, which only the rich could afford, acted as an expression of membership in the upper class. Nowadays, thanks to fast fashion, off-price stores, a global market place, technological improvements, it may be harder to discern luxury items from similar products and lower price points.

3.2   Market research

3.2.1   Definitions

A **market** is a place for buying and selling, for exchanging goods and services, usually for money. When developing new products, the key is to bear the consumers in mind. However, the demands, motivations, lifestyle and product choices of consumers are constantly evolving. By analyzing and conducting market research, a company can better identify which customer groups will purchase its product or service and establish the brand's target market.

Market research is an organized effort to gather information about target markets or customers. Market research investigates what products the target market wants and how to make the crucial link between the product and the market. It also considers the past, present and potential customers for the product. Market research reviews the characteristics, spending habits, location, and consumer needs, the overall industry, and particular brand competitors.

3.2.2   Market research process

Research procedures may vary depending on the nature of the research problem, but in general, the process of market research consists of a number of stages: formulate the problem, design the research, test the research design (pilot), collect the data, analyze the data and interpret the data, and present the findings.

Formulating the problem

The process begins when someone in the organization sees a need for information. For example, a small-business owner might wonder whether his or her product was selling better in department stores or in specialty shops. When a marketer needs information, he or she must formulate the problem. Defining the research problem is the most critical step in the research process. Unless the problem is accurately defined, the information collected will be of limited or no use.

Design the research

Research design is the plan for how to collect and analyze the data. At this stage, researchers need to determine data sources, select the sampling method, select the data collection method, and probably design the data collection form (questionnaire). Based on the problem previously defined, the researcher selects one or more of the basic research designs: exploratory, descriptive and causal research.

Exploratory research gathers information from whatever sources are likely to provide useful insights. When researchers seek to discover ideas and insights, they are conducting exploratory research. This is most useful in the early stages of research, particularly if the researcher is not familiar with the subject area.

Descriptive research studies how often something occurs or what, if any, relationship exists between two variables. It will provide an accurate description of the variables uncovered by the exploratory stage. This research could be used to investigate the market share of a company's products or the demographic characteristics of the target market (age, gender, income, etc.). Data are usually obtained from secondary data sources or from surveys.

Causal research looks for cause-and-effect relationships. For instance, it doesn't just investigate whether there is a relationship between advertising and repeat purchases, rather it seeks to find whether the advertising causes people to choose the product.

3.2.3 Collecting data

Market research may be quantitative or qualitative. **Quantitative research** involves objective methodology in which data are collected about a sample population and analyzed to generalize behavioral patterns. **Qualitative research** is more subjective and relies on methodologies such as observation and case studies in which experiences are recorded as a narrative to describe observed behaviors within the context of environmental factors.

Information needed

Marketers need information about the target markets, consumers and competitors for marketing planning. The marketing mix is the combination of elements that a fashion marketer offers to a target market. It comprises decisions made about products, prices, promotion, and distribution (placement) that are assembled in a coherent manner to represent the firm's offering to the consumer. Table 3.2 lists the key information for market research corresponding to the 5Ps+C.

Table 3.2　Key information for market research

5Ps+C	Questions to ask
People	Who are the current and potential customers? Where do they live? How much do they earn?
Product	What are the benefits being delivered to the customer? Which design is likely to be successful? What kind of packaging should be used?
Placement	Where is the product available? What type of outlets carries the product? What are the dominant channels of distribution?
Price	What price should be charged for the products?
Promotion	What is the core selling proposition, and how can it be communicated most effectively to customers: advertising, public relations, direct sales? How can special offers incrementally induce or accelerate sales?
Competition	How sensitive is the market to price? What is the elasticity of price movements up or down? Are there distinct price segments? Is the brand gaining or losing share at the current price level?

Sources of information

Data come from two sources, primary and secondary. Secondary sources consist of information that has already been collected for other purposes and primary sources of information are those used for the purpose of collecting information specifically for the current research project.

　　Secondary sources. Secondary sources can be separated into two types of internal and external. Internal sources are those that generate information within a company or an organization, e.g. sales figures, company reports, customer complaints and accounts information. External sources are those that generate information outside a company or an organization. They can be government statistics, trade information, financial reports, fashion/trade publications and online websites such as *Women's Wear Daily* (*WWD*) and *World Global Style Network* (*WGSN*).

　　Primary sources. Primary information can be from consumers, and the fashion industry (designers, retail buyers, manufacturers, retailers), and so on.

3.2.4　Data collection methods

Primary data collection methods

The main approaches to primary data collection are observation, focus groups, interview and surveys.

Observation. Sometimes people don't do exactly what they say they will, and sometimes users of a product can't fully describe their experience with it. For these reasons, researchers may use observation, or the collection of data by recording actions of consumers or events in the marketplace. For example, in-store video cameras may be used to record behavior, for analysis such as researching store layout.

Focus group interview. A focus group interview is a personal interview of a small group of people in which the interview poses open-ended questions and encourages group interaction. The group usually consists of between 6 and 12 respondents who discuss products, services, attitudes or other aspects of the marketing process. The discussion is led by a skilled researcher called a group moderator, who guides the discussion, following a checklist of topics. These discussions can take several hours to complete and are often used as a preliminary to survey research. Nowadays, it is possible to conduct online focus groups via the Internet.

Surveys. Surveys involve the systematic gathering of information from respondents by communicating with them in person, over the phone, by email, or through the Internet. The information covers such topics as consumer behavior, attitudes and beliefs, as well as buying intentions. The strengths of these beliefs, attitudes and intentions are measured and the results are extrapolated to the population as a whole. Survey research is widely used for descriptive research.

Questionnaire design

The questionnaire is a vital part of most surveys and great care must be taken with its design. Many factors will affect the design of the questionnaire, such as the nature of the data required (qualitative or quantitative) and how the questionnaire is to be administered (by personal interview, telephone, mail or other self-completion, or whether electronic instruments will be used). A well-designed questionnaire will provide the researcher with complete, accurate and unbiased information using the minimum number of questions and allowing the maximum number of successfully completed interviews.

3.2.5   The Internet as a research tool

The increased use of the Internet has had a great impact on the market research. Online commercial databases have been available to researchers for many years, providing access to news sources, trade publications and market reports. Collection of primary data via the Internet by e-mail or website-based surveys or by online discussion groups become faster and easier. Conducting research via the Internet may increase the speed of research from design to results and reduce costs, as well as appearing to facilitate research on an international level.

3.3   Market segmentation

3.3.1   Need for market segmentation

Each consumer is unique. All consumers are different from other consumers, but they are also similar to some other consumers. Besides bespoke tailoring and couture items, mass fashion market provides standardized garments aimed at particular groups of consumers. Fashion firms demand that groups of consumers with similar needs be identified and then supplied with similar products.

Market segmentation is where the larger market is heterogeneous and can be broken down into smaller units that are similar in character. In practice there is always the problem of balancing the similarity of needs with the desire for substantial numbers of potential buyers. Formally defined, market segmentation is the process of subdividing a market into distinct subsets of customers that behave in the same way or have similar needs. The purpose of market segmentation is to identify what subset of people will potentially buy the product, and to define the **niche market** for the product.

3.3.2   Approaches to segmenting consumer markets

There are two basic categories of segmentation used for consumer markets: segmentation based on characteristics that describe consumers, and segmentation based on consumers' relationship to the product.

Segmentation based on descriptors

One approach to segmenting markets is to describe the characteristics of potential customers. Such descriptions tend to look at demographic, geographical or psychographic characteristics of the buyer or a combination of the three measures.

Demographic segmentation. Demographics are statistics about a given population with respect to age, gender, marital status, family size, income, spending habits, occupation, education, religion, ethnicity, and region. Demographic segmentation is dividing the market based on demographics.

Geographic segmentation. Geographic segmentation divides the market into groups according to a specific region. It is deemed that people who live in the same area may share similar needs and wants compared to people who live in other part of the country. Here are some examples of geographic variables.

• Region: by continent, country, city, suburb or neighborhood;

• City size: segmented according to size of the population;

• Population density: urban, suburban, rural;

• Climate: various weather patterns are common to certain parts of a region, like hot, temperate, or cold.

Psychographic segmentation. Demographic data fail to capture the activities, interests, and opinions that differentiate populations within statistical groupings. **Psychographics** is the study of the social and psychological factors that influence consumer lifestyles; these data reveal motivations and buying practices. Therefore, psychographic segmentation groups consumers based on their activities, interest, and opinions by looking at these data.

Segmentation based on benefits and buying behavior

Another approach considers consumers' relationship to the product: what benefits they are looking for, how much they buy, how loyal they are to a particular brand.

Benefit segmentation divides a market according to the kinds of benefits desired. Researchers gather information of the values and perceptions of consumers through surveys.

Segmentation based on consumer behavior divides the market on the basis of product purchases, for example into heavy, moderate, and light users. One measure of whether consumers might be heavy users of a particular brand is their brand loyalty. Brand loyalty refers to the consistency with which a consumer continues to buy the same brand of a particular product.

3.3.3 Defining your target customer

Select a target market

A business using market segmentation would concentrate its efforts on the market segments it can serve profitably. The market segments on which an organization focuses its efforts are known as its **target markets, or target segments**. There is no one best way to select a target market. There are some main criteria that should be taken into account to achieve maximum profits or effectiveness:

- The segment should be measurable and easily identifiable. Producing garments to appeal to "art lovers" may sound a good idea, but until it is known how to identify "art lovers" and work out the market size, little progress can be made;
- The chosen segment or segments should be relatively stable. While styles may evolve over successive seasons, it is hoped that the core segment will remain loyal to the target consumers;
- The segment or segments chosen should be accessible. Some target markets are difficult to reach at a reasonable cost. As mentioned early, demographic data help to simplify decisions about reaching the chosen segments;
- The segment should be large enough to be profitable for the scale of the organization. Depending on the organization's product, capacity and access to resources, a bigger segment may be more attractive than a small one.

Create a customer profile

Knowledge of the customer is a critical prerequisite to the design process. Once the target markets have been identified, the attitudes and perceptions of potential consumers may be researched, and a customer profile or pen portrait for each one is created. A customer profile is a written description of the kind of person that the fashion retailer is selling to and includes many of the segmentation variables. A pen portrait typically includes an age range and price point that the customer is willing to pay based on demographics. It also identifies general lifestyle characteristics and their shopping point of view or behavior. An example of a customer profile is given below (take Ralph Lauren as example).

Ralph Lauren

Demographic: a variety of different demographics that include women, men and children of all ages.

Geographic: the United States, Canada, Europe, Japan, South Korea, China, Southeast Asia, and Latin America. The Americas account for about two-thirds of sales, while Europe and Asia generate 21% and 12%, respectively.

Psychographic: family oriented, values the simple aspects of life, more elegant outdoor sports such as polo and rugby, dress more formally with a classic, elegant, and sophisticated style, take pride in looking, put together in a very classic and every-day formal way.

◎ Summary

The fashion consumer decision is a comprehensive process and consumer behavior is influenced by many individual and sociological factors. Market research is essential in fashion product development. There are many methods to collect primary and secondary data from various sources for market research. Consumer market can be segmented based on either demographic, geographical or psychographic descriptors, or benefits and buying behavior. Some main criteria are often used to select a target market.

◎ Key terms

Athleisure 运动休闲风   Demographics 人口信息

Cognitive 认知的   Generational cohort 时代群体

Customer profile 客户档案   Likert scale 利开特式量表

Decision-making 决策   Marketer 市场营销者

Market research 市场调研

Market segmentation 市场细分

Masstige 奢侈大众化

Niche market 缝隙市场，小众市场

Personality 个性

Psychographics 心理统计特征

Purchase 购买

Qualitative research 定性研究

Quantitative research 定量研究

Questionnaire 问卷

Subculture 亚文化

Surveys 调查

◎ Product development team members

The following product development team members were introduced in this chapter:

Marketer（营销人员）: an individual engages in various activities of fashion marketing. His or her responsibilities include market research, promotion, product positioning and pricing, and distribution.

Market researcher（市场研究者）: an individual engages in market research. He or she may investigate the market shares of competitors and trends in those shares.

◎ Activities

1. In class, identify several brands that do an excellent job of relating to you as a consumer, discuss the strategies and product characteristics that make each brand successful.

2. Through group discussion, identify some general consumer trends.

3. Design a store observation study to better understand consumer preferences. Report your observations and insights in class.

◎ Review or discussion questions

1. Discuss your priorities in shopping for apparel and differentiate those priorities from those of your parents. How do your perceptions of value differ? Distinguish between the types of shopping environments and advertising messages that appeal to you and those that appeal to your parents.

2. How have companies solicited your consumer input in the past? Has this chapter made you more aware of information that is being collected without your realizing it?

◎ Semester project

Semester project I : Business strategy report—Part 2

Define your target consumer:

Create a Customer Profile Worksheet for your brand.

◎ References

[1] EASEY M. Fashion Marketing[M]. New Jersey: Wiley-Blackwell, A John Wiley & Sons, Ltd., 2009.

[2] KEISER S, VANDERMAR D, GARNER M B. Beyond Design, the Synergy of Apparel Product Development[M]. 4th Edition. New York: Fairchild Books, 2017.

[3] BURKE S. Fashion Entrepreneur-Starting You Own Fashion Business[M]. 2nd Edition. New York: Burke Publishing, 2013.

[4] REGAN C L. Apparel Product Design & Merchandising Strategies[M]. New Jersey: Pearson Prentice Hall, 2008.

[5] RATH P M, BAY S, PETRIZZI R, et al. The Why of the Buy: Consumer Behavior and Fashion Marketing[M]. New York: Fairchild Books, 2008.

[6] CHURCHILL G A, PETER J P. Marketing: Creating Value for Customers[M]. Austen Press, 1995.

[7] JACKSON T, DAVID S. Mastering Fashion Buying and Merchandising Management[M]. New York: Palgrave Macmillan, 2001.

◎ Useful websites

[1] Retail customer experience, www.retailcustomerexperience.com.

[2] Women's Wear Daily (WWD), www.wwd.com.

[3] World Global Style Network (WGSN), www.wgsn.com.

[4] ESRI, www.esri.com/data/community_data/community-tapestry/index.html.

[5] About Retail, www.retailindustry.about.com.

[6] Internet retail, www.internetretailer.com.

[7] Ralph Lauren,https://ralphlaurenbrandanalysis.weebly.com/target-consumer.html.

[8] Retail bulletin, www.theretailbulletin.com.

[9] Retailing today, www.retailingtoday.com.

[10] Retailwire, www.retailwire.com.

◎ Appendix

Table A3.1 Generational cohort groups active in today's marketplace

Generation	Born	Shared experience
Matures, includes Gls (Civic) and Silents (Adaptive)	1909–1945	Great depression, the World War Ⅱ, Korean War, Cold war; Great civic organizers
Baby Boomers (Idealist)	1946–1964	Born during postwar expansion and prosperity, unprecedented employment and educational opportunities; social activists
Generation X (Reactive)	1965–1978	Born in the wake of tumultuous political and economic conditions; came of age during a period of business downsizing and reengineering; many were products of broken homes; grew up with MTV and the AIDS epidemic
Millennials or Generation Y (Civic)	1979–1995	Raised during the longest bull market in history; faced a war on terrorism and a recession; more diverse than previous generations and more tolerant of diversity; schooled with technology
Centennials or Generation Z (Adaptive)	1996–present	Born and grew up during an era of terrorist threats, extended economic and political gridlock, and increasing awareness of the fragility of our environment and natural resources. More pragmatic about the future and more practical in their purchasing habits. Have been raised with technology

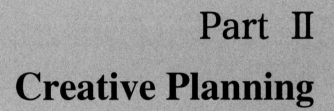

Part Ⅱ
Creative Planning

Chapter 4

Trend Forecasting

课题名称：Trend Forecasting

课题内容：时尚趋势预测

课题时间：4 课时

教学目的：

1. 更全方面地理解时尚类别并了解时尚趋势的生命周期。

2. 学习利用环境分析来识别长期趋势和短期趋势。

3. 学习识别各种资源来完成趋势预测。

4. 理解市场采购的重要性。

5. 熟悉色彩、面料和廓形的各种信息，完成趋势预测。

6. 理解特定市场对趋势的解读。

7. 了解趋势预测部门的工作。

8. 熟悉各季节趋势预测的设计版式。

4.1 What is fashion?

4.1.1 Definition of fashion

Fashion is a popular aesthetic expression, especially in clothing, footwear, lifestyle, accessories, make-up, hairstyle and body proportions. There are many dimensions existing in fashion. It is a term which is defined with different levels, such as artistic, functional, scientific, and business aspects. Fashion is a communication tool which can present the affiliations and roles of the wearer in the society. It indicates the age, gender, power, and sexuality of a wearer. Fashion is evolutionary. It moves and morphs through a cycle of popularity which is the essence of fashion.

A trend often connotes a very specific aesthetic expression, and often lasts shorter than a season. Fashion is a distinctive and industry-supported expression traditionally tied to the fashion season and collections. Trend forecasters must understand the dimensions of fashion in order to interpret trends for their specific market.

4.1.2 Different levels of fashion

There are generally four levels of fashion: Paris couture, designer ready-to-wear, street fashion, and functional fashion.

Paris couture

Garments are characterized by luxurious fabrics, complex silhouettes, meticulous tailoring, exquisite beading, and unique details. In this level, designers focus on artistic and creative expression, along with a quest for publicity, rather than practicality, function, or profit motives.

Designer ready–to–wear

Designer ready-to-wear styles may be produced in quantities that vary from 100 garments to several thousand. Although they are not as expensive as couture garments, they are beautifully designed, impeccably made, and use the finest fabrics.

Street fashion

Street fashion originates with the consumer rather than a designer or product developer. Free spirits and innovative youth express their creativity by putting different eclectic but inexpensive sources together for looks that define who they are and how they think.

Functional fashion

Functional fashion takes advantage of scientific developments in textiles, fabric construction and finishing, and electronics to make high-tech apparel that appeals to the needs of niche markets.

4.1.3   The fashion circle

Trend

A trend is the preference for a particular set of product characteristics within a consumer group. Trends may refer to innovations in fiber and textile, or the popularity of a particular color, silhouette, or garment detail.

Fashion circle

Fashion is defined by trends. Fashion circle presents the acceptance and rejection of fashion trends. Fashion circle tracks the movement, pace and direction of fashion trends. Normally, fashion circle is represented by a bell-shaped curve that is plotted by using a vertical axis that represents unit sales and a horizontal axis that represents time. As shown in Figure 4.1, fashion trends evolve through an introductory phase, to mass market saturation, and finally into obsolescence.

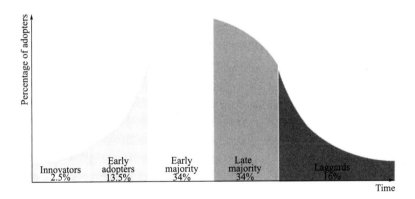

Figure 4.1   The fashion cycle

Functions of the fashion cycle

The fashion cycle plays an important role in the decision-making in designer signature stores, design departments in department stores, and high-end contemporary boutiques. With the utilization of the fashion cycle, product developers are able to know when their target customers are ready to purchase and also to be ready to offer the latest trends during the introduction phase.

4.2   Environmental scanning

4.2.1   What is environmental scanning?

Environmental scanning is the ongoing process of surveying a variety of resources for economic, political, social, technological, and cultural conditions for insights into the future.

4.2.2　Why should we carry out environment scanning?

Through environment scanning, designers, trend specialists, and merchandisers are able to anticipate the resulting impact on fashion and lifestyles when the environment changes. It reduces the risks inherent in trend forecasting. It recognizes a trend on its upward ascent and offers merchandise that is well-timed.

4.2.3　Different types of forecasting

Long–term forecasting

Long-term forecasting is the process of analyzing the sources, patterns, and causes of environmental change through the evaluation of current events in order to identify and anticipate directional shifts in business strategies, consumer behavior and lifestyle, and global dynamics.

It seeks to identify:

- Major shifts in domestic and international demographics;
- Changes in business, industry, and market structures;
- Changes in consumer interests, values, motivation, and circumstances;
- Breakthroughs in technology and science;
- Changes in the domestic or global economic picture;
- Shifts in political, cultural, or economic alliances between countries.

Short–term forecasting

Short-term forecasting focuses on current and upcoming events, and pop culture phenomena that can be translated into fashion trends in the next 12 to 18 months to give fashion a fresh look each season.

Environmental scanning relies on an analysis of news in a variety of categories. These influences affect both business trends and fashion trends.

4.2.4　Environmental scanning influences

Current events

Global current events have a major impact on fashion. All the continued economic and political uncertainty around the world and the ongoing environmental issues have caused great changes in the consumers' preferences and have inspired the designer on fashion.

The arts

The arts have a major impact on fashion. Designers and fashion forecasters make a point of being the first to see a major art opening or historic costume exhibition. Color forecasters respond to the mere announcement of a major art exhibition. An exhibit of painting by Monet, Van Gogh, Matisse, or Klimt or popular culture like movies, is sure to influence both seasonal color palettes and textile patterns.

Sports

Popular sports frequently influence fashion. Sports celebrities can give athletic lines tremendous appeal. Sports can also be a driver of textile innovation.

Science & technology

Science and technology affect many aspects of fashion, from the colors we can achieve on different mediums, to the fabrications available to designers, to how garments function, to how we care for and dispose of garments.

4.2.5   Environmental scanning resources

Environmental scanning requires collecting as much information as possible to get trends right.

Professional media

- **National newspapers:** *New York Times, Washington Post*;
- **Consumer trends:** *Advertising Age, Brandweek*, www.trendwatching.com;
- **Technological developments:** *Fast Company, Wired, Scientific American*;
- **Publications covering the fine arts:** *Art and Antiques, Vanity Fair, Ornament, Architectural Digest*;
- **Apparel industry/textile/fashion news:** *Collezioni Donn, Vogue, Collezioni Trends, Textile View*;
- **Internet resources:**www.style.com, www.vintagefashionguild.org, www.hintmag.com, www.wgsn.com, www.wwd.com, www.fashion-era.com, www.fashion.about.com, www. businessoffashion.com, www.madmuseum.org, www.trendtablet.com, www.fashion-incubator. com, www.coolhunting.com, www.vogue.com, www.elle.com;
- **Famous blogs:** *The Sartorialist, The Fashion Guitar, Wish Wish Wish, Style Bubble, Bryanboy, Menswear Style, Man Repeller, Camille Over the Rainbow, Advanced Style, Tommy Ton, Pandora Sykes, The Blonde Salad, Song of Style, This Time Tomorrow, That's Chic, Wendy's Lookbook, Gary Pepper Girl, Helena Bordon, Tanya Burr, The Chronicles of Her.*

Social media

Instagram is a free social application software provided by Facebook to share online pictures and videos. It allows users to take photos with smartphones, then add different filter effects to photos, and then share them with social network services. They can be seen and commented by anyone all over the world.

Weibo is a kind of broadcast social media and network platform based on user relationship information sharing, dissemination and acquisition, which shares short and real-time information through an attention mechanism. In 2021, data shows that the number of monthly active users reached 573 million. Weibo presents a media character with wonderful real-time performances. Information spreads rapidly and extensively.

WeChat provides functions such as public platform, friend circle, message push, etc. Users can add friends and pay attention to the public platform by "shaking" "search number" "nearby people", and scanning QR code. Meanwhile, WeChat shares the content to friends and the wonderful content users see to WeChat friend circle. Fashion information can spread rapidly through chains and also focus on different groups.

Red book is a lifestyle platform and consumption decision-making entrance which provides a chance for those fashion buyers who promote their stuff and popularize fashion information. There are some similarities between Weibo and Red book while Red book is a community and it' s more concise. The users share notes through texts, pictures and videos and most of them are of young age. In August 2014, Red book launched the e-commerce platform "welfare society", which upgrades e-commerce from the community and completes the business closed-loop (Tables 4.1 and 4.2).

Table 4.1　Different common social media and their functions & content & audience

Social media	Functions	Content	Audience
Weibo	Media Broadcasting & Transmit	News & Information	All users
WeChat	Delivering & Transmit	Articles & Information	Readers & Friends
Red book	Users' sharing	Instruction & Information & Notes	All users

Table 4.2　Sources for media scanning in different subjects

Subject	Source
General news	Televised/cable network news: *ABC, CBS, CNN, FOX, MSNBC, NBC*, and *PBS* News weeklies: *US News* and *World Report and Time* National newspapers: *New York Times, Washington Post*, and *USA Today*
Business, consumer, and technology trends	Consumer trends: *Advertising Age, Brandweek,* www.trendwatching.com, www.firstmatter.com,

continued

Subject	Source
Business, consumer, and technology trends	www.iconoculture.com, www.trendsearch.com, Business news: *Business Week, Forbes, Fortune, Wall Street Journal* Technological developments: *Fast Company, Wired, Scientific American*
Publications covering the fine arts, performing arts, and popular culture scene	*Architectural Digest, Veranda, Art and Antiques, Ornament, American Art Review, Vanity Fair, New Yorker*
International views	*The Economist, Al Jazeera, BBC News, The Guardian*
Apparel industry/textile/fashion news	WWD.com, just-style.com, *Collezioni Donna, French Vogue, Collezioni Trends, Textile View, View 2, Fashion Trends Forecast, NRF Smartbrief,* vogue.com, elle.com, sourcing journalonline.com Specific markets: *Earnshaw's Review, Accessories, Kidswear, Sport & Street,* sportswear-international.com

4.3   Shopping the market

4.3.1   Why do shopping in the market?

After the trends are identified through environmental scanning, product developers and trend forecasters must verify and then interpret them for their target customers. And shopping the market makes it happen. Trend forecasters rely on shopping the market to get a comparably consistent preference of the customer on how the trends are interpreted. Through shopping the market, they get to know how product of their own is being merchandised—where it's selling best and where it's not resonating the customers. They must also shop the competition to understand how their competitors have interpreted the same trend and how consumers have responded, helping them to make adjustments to the product or the design.

4.3.2   Determining where to shop

Where to shop depends on which shopping venues will yield the most valuable information for their specific market, product category, price point, and fashion calendar. If they develop outerwear, they might shop in Toronto; for a resort, they may visit Saint-Tropez. New York, Los Angeles, Paris, Milan, Barcelona, Tokyo, and Antwerp are other common shopping destinations. Product developers often have large budgets to buy samples that will become the basis for styles in their seasonal lines while the designers tend to purchase garments only when a material or construction is so unusual that they require it for reference.

4.4　Dimensions of fashion trends

4.4.1　Color

Why color matters?

Color is one of the first stimuli a customer responds to when shopping. That means it's a key to the trend forecasting. The decision about a seasonal color palette is one of the first to be made in the product development process. Color forecasts need to be interpreted for different consumers groups and specific market categories.

The color forecasting process

The color forecasting process begins two to two and a half years in advance of a selling season. It is based on environmental scanning, which identifies the non-fashion events that influence fashion trends and lifestyle themes.

Color associations

International Commission for Color in Fashion and Textiles. The International Commission for Color in Fashion and Textiles founded in 1963 is a non-profit organization and a leading organization in the field of international color trend. At present, it is the most authoritative organization that affects the world's fashion colors of clothing and textile fabrics. The international popular color committee holds two-color expert meetings every year to formulate and launch four groups of international popular color cards for men's and women's wear in spring, summer and autumn and winter, and put forward the color inspiration and sentiment of the popular color theme, so as to provide new inspiration for the color design of fashion and fabric.

　　China Fashion Color Association. Established in 1982 with the approval of the Ministry of Civil Affairs of the People's Republic of China, China Fashion Color Association is a legal person social organization composed of institutions and personnel engaged in research, prediction, design and application of pop colors. In 1983, it joined the International Commission for Color in Fashion and Textiles on behalf of China. As a national association directly under China Association of Science and Technology, it is affiliated with China Textile Industry Association.

Color forecasting based on materials (textile consortiums)

Consortiums of textile manufacturers develop seasonal color stories which are geared to the end-use categories of the markets they supply. The predictions come to life as textile manufacturers present their seasonal lines at global fabric fairs that occur about one year before a consumer season.

Color forecaster

Color forecaster Alison believes that color cycles can be tracked using a bell curve similar to the one used to measure fashion cycles. Figure 4.2 shows a color cycle explaining how colors are introduced, increase in popularity, their use becomes saturated, and then they become obsolete, typically over a three-year period.

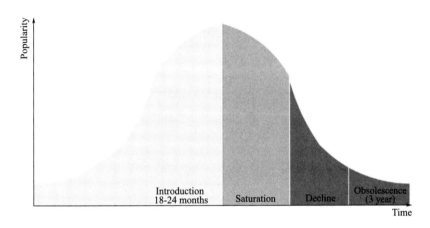

Figure 4.2   The color cycle

4.4.2   Fabric

Timing of fabric trend forecasting

The timing of fabric trend forecasting goes on simultaneously with color forecasting.

Why fabric trend forecasting matters?

The fabric forecasts sources alert the product developer to new technology, fibers, blends, and finishes by providing descriptions, swatches, and sketches of possible applications. Designers may rely on these sources and on magazine sources for preliminary research to recognize new materials in the marketplace.

Fabric shows

Fabric is a medium that must be touched and draped to be appreciated. Toward that end, a number of domestic and international fabric and yarn shows are held each year to give product developers an overview of what is available. At a fabric show, the yarn and fabric vendors will introduce their seasonal lines and offer their prototype garments to help sell their newest products, and the designers and product developers will confirm developing trends, identify new resources, and order sample fabric yardage which can be used for creative experimentation before committing to production yardage.

Important shows include the Première Vision and Texworld, both in Paris and New York, and Kingpins and Asian fabric exhibits.

Fabric purchasing

Most of the buying that takes place at large fabric and fiber shows is for sample quantities. Larger commitments are made after the product developer has had time to experiment with the fabric and determine how important it will be to the seasonal line to minimize risk.

Trend of printed fabrics

Printed fabrics can make some certain categories product developers' line unique. Children's apparel, dresses, and lingerie are examples of categories that rely heavily on prints.

4.4.3　Silhouette

Why silhouette matters?

Silhouette is a term used to describe the outline or shape of a garment. The interpretation of silhouette will vary from shape, fabric weight, placement on the figure, and proportion, but having a defined silhouette gives the collection a sense of cohesion.

Generation of silhouette inspiration

Silhouette inspiration comes from a variety of sources. Suggested by trend services, silhouettes and details are usually changed for some specific aesthetic or markets needs as the fashion sketches are modified correspondingly. Since more and more fashion ideas trickle up from the streets rather than down from the runways, shopping the market is perhaps the best source of silhouette inspiration.

4.5　Seasonal trend forecasting

4.5.1　What is season?

Each seasonal collection is the result offered by a product developer of trend research focused on the target market it has defined for itself. Product developers may offer anywhere from two to six seasonal collections per year, depending on the impact of fashion trends in a particular product category and price point (Table 4.3).

Table 4.3　Discussion on women's wear/men's wear/children's wear

Garment category	Seasonal collections per year
Women's wear	4-6
Men's wear	2-4
Children's wear	3-4

4.5.2　Who is responsible for trend forecasting?

Some large companies have their trend departments that are responsible for pinpointing the initial trends of the season for their company. Merchandisers who come from a buying background and are skilled at anticipating what the customer is ready for are also key players in determining trend direction. Afterwards designers translate the trends identified into saleable garments in the right colors, fabrics, and silhouettes. In the end, it's the collective vision of these three groups that determines a given brand's direction.

In companies that develop wholesale brands, the design team and the merchandiser generally work in tandem.

4.5.3　Forecast formats

The members of the product development team responsible for trend forecasting use a variety of formats to share their seasonal research. The degree of formality in the presentation depends on how widely within the organization the information will be presented.

Designers for wholesale brands are responsible for both the research and creation of the line. They tend to work directly from informal collections of swatches and tear sheets to select the actual colors, fabrics, and prints they will use for each item or group in the line.

Trend teams for private brands must prepare more formal presentations because their forecasts are applied to multiple categories across the men's, women's, and children's divisions. In these organizations, a few key members are responsible for interpreting the appropriate trends for each division and category that develops product (merchandisers, creative and technical designers) and the buying staff. Finally, product developers sell the line they develop to the buying team. When the line is finalized, early collaborations make the process where the buyers are better prepared to commit to the quantities more smoothly.

Then trend forecasters frequently develop a seasonal trend book that highlights important trends in color, fabrics, silhouette and details for each category and division served. Each team receives its own trend book, development samples are hung around the room, story boards are displayed, and a PowerPoint presentation summaries the most important trends for the season.

◎ Summary

Fashion is a reflection of our times, a means of expression that reflects how we respond to everything happening within our own lives and in the world around us. Fashion is available at different levels. Fashion is influenced not only by high fashion but also by developments in functional apparel and street fashion. Fashion trends evolve in cycles, with fads lasting for a very short time and fashion classics lasting for many seasons. Product developers at all levels of fashion use environmental scanning to identify trends that will have an impact on consumer lifestyles and market-place conditions. Shopping the market helps trend forecasters see how trends are playing out in fashion centers around the world.

◎ Key terms

Couture 高级时装、高级定制

Street fashion 街头时尚

Fashion show 时装秀

Functional fashion 功能性时装

Environmental scanning 环境分析

Long-term forecasting 长期预测

Ready-to-wear 成衣

Short-term forecasting 短期预测

Silhouette 廓形

Trend 趋势

◎ Product development team members

The following product development team members were introduced in this chapter:

Trend department（趋势部门）: people who are responsible for pinpointing the initial trends of the season for their company.

Trend forecasters（趋势预测者）: people who search for facts and then analyze the findings to predict trends that will positively affect the amount and types of fashion products consumers will buy.

Merchandiser（营销师）: people who are skilled at anticipating what the customer is ready for and are also key players in determining trend direction.

◎ Activities

Using the Internet and recent periodicals, identify several trends recently promoted by futurists. Brainstorm as to how these trends will influence fashion. Design students should take one prediction and design a series of garments inspired by that prediction in their sketchbooks.

◎ Review or discussion questions

1. In class, look at some photos from the most recent couture or designer collections. Which elements in each design might go on to influence mainstream fashion?

2. Identify a current fashion that is likely to be short-lived. Identify a current fashion that appears to be a fashion classic. How do they differ?

3. Identify several current events in the political, global, and cultural arena. What impact might they have on fashion?

4. What colors are currently popular in various consumer markets? Try to distinguish color preferences at several different price points and in several specialty markets.

5. How has Burberry used its signature plaid in new and innovative ways to revitalize the popularity of the label?

◎ Semester project

Semester project Ⅱ: Create a trend board—Part 1

1. Find your inspiration. You have to find what inspires you. Maybe it's travel, flowers, art, architecture, but it can be whatever you want.

2. A clear color palette, making sure that the fabrics, prints, materials are tied in together. You can also include a "color standard", which might be a Pantone code, or a swatch of fabric.

3. Do "primary research": go out + take photos of something you find inspiring.

4. Do "secondary research", which can be the ideas found in books + online that relate to your theme + also trend research.

5. Put your trend board together. To make it finished, you can use apps like Adobe Photoshop, Moodboard on the go, etc.

Example (Figure 4.3):

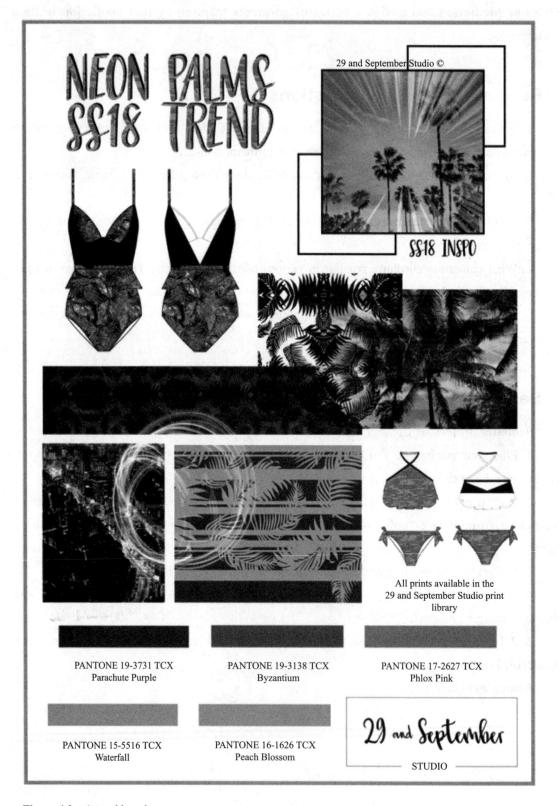

Figure 4.3   A trend board

◎ References

[1] BRANNON E L. Fashion Forecasting[M]. New York: Fairchild, 2005.

[2] LURIE A. The Language of Clothes[M]. New York: Vintage Books, 1981.

[3] PORTER T. Color in the Looking Glass[M]. In Color Forecasting, ed. LINTON H, 1-9. New York: Van Nostrand Reinhold, 1994.

[4] MILESS. The Paris couture: A crop of fresh faces suggests new vitality[J]. Women's Wear Daily, July 8, 2002.

[5] WATERS R. The Hummer and the Mini: Navigating the Contradictions of the New trend Landscape[M]. New York: Penguin Group, 2006.

[6] WEBB A L. Timing is Everything[M]. In Color Forecasting, ed. LINTON H, 203-206. New York: Van Nostrand Reinhold, 1994.

◎ Useful websites

[1] Just style. www.just-style.com.

[2] The Fashion Business Coach. www.thefashionbusinesscoach.com.

◎ Case study: WGSN (trend forecasting)

WGSN (formerly Worth Global Style Network) is a trend forecasting company of parent organization Ascential. WGSN was founded in 1998 in West London by brothers Julian and Marc Worth. Emap (now Ascential), a business-to-business publisher and exhibitions company, bought the company in October 2005. They constantly monitor the signals of change that will impact how consumers think, feel and behave. Their experts connect the dots to accurately predict the products, experiences and services people will need in years to come, helping brands stay relevant and secure their place in the future. Make confident decisions backed by trusted trend insights, expertly curated data and actionable recommendations, from up-to-the-minute trend updates to 10-year-out forecasts.

They make it easy for creative thinkers to understand and design for the next-generation consumer, wherever in the world you might be based. They scan the global landscape so we can focus on creating outstanding products and experiences.

Their extensive network of industry experts, based in 15 regional offices, provides inspiration to a community of over 43,000 product designers and thought leaders in 32 global markets. Design the products, services and experiences that people will want to buy by tapping into the most trusted consumer insight and product design direction. The data-backed insights and custom advisory solutions help you make accurate decisions, stay ahead of change and know how to adapt.

For over 20 years, WGSN has been getting it right in trend forecasting, powering the world's most valuable brands. Launched in London in 1998, WGSN disrupted the market with a pioneering online trend service back then, and it continues to drive the industry by forecasting change.

Chapter 5

Color Management

课题名称：Color Management

课题内容：时尚色彩规划

课题时间：2课时

教学目的：

1. 认识色彩的物理原理。

2. 认识色彩决定在整个供应链中的影响。

3. 明确色彩规划的专业术语。

4. 概述视觉和数字的色彩审批过程。

5. 识别色彩规划的影响因素。

6. 分析在色彩测试中科技的影响力。

5.1 Seasonal color palette considerations

5.1.1 Why color matters?

Color draws customers in and entices them to a rack in order to explore other aspects of the garment, including fabric and silhouette. Color decisions are based on both consumer demands and production efficiencies. Product developers strive to offer enough choices to stimulate a broad base of customer interest without offering so many choices that the customer becomes confused. Developers must also consider how their color assortment affects manufacturing.

5.1.2 Seasonal color

Seasonal color is a palette of colors traditionally associated with the different selling seasons or deliveries, to signal a change of season and to entice consumers to buy. Variations of color will occur based on trends. The designer price point and haute-couture markets often ignore traditional color palettes to get their own vision for the collection, to supply trend information to lower-priced markets. Colors that are often used as seasonal colors, not necessarily all in the same collection (Table 5.1).

Table 5.1 Seasonal color

Season	Colors
Transition	Brown, olive green, pumpkin, cranberry, ocher yellow, dark khaki, charcoal gray, Taupe, black, chocolate brown, deep rich colors
Holiday	Metallics such as silver, gold, and bronze; champagne; ivory, black; jewel tones such as sapphire blue, ruby red, emerald green
Resort/pre-spring	Soft pastels, white, navy blue, cherry red, bright green, tan
Spring	Bright colors such as yellow, Kelly green, indigo, purple, navy, light khaki
Summer	White, bright saturated colors

5.1.3 The complexity in color story design

Each size and color in which a style is offered represents a stock-keeping unit (SKU). The more SKUs there are to manage, the more complex color management and its impact on sourcing and buying decisions become. Every color that is used in solid, yarn-dyed, or patterned fabric must be included in the seasonal color story. The same color used in several different fabrications must be managed to match throughout the line; the notions used with each style must also be color-matched.

5.1.4   Merchandising considerations in color management

Customer

Color that is worn becomes part of the wearer's physicality and personality.

Color scale

Color is used sometimes it isn't the color that seems inappropriate, but the contexts surrounding it.

Define the color relationships

Define these colors and tones first to provide a foundation for all others, when considering which one core fabric will create the foundations of your collection.

The matic and contextual representation

How colors are extrapolated from a mood board to create a palette that conveys the desired feeling relies heavily on their context.

The "middleman"

It is essential to create a middleman to unify unrelated and disjointed solid colors in a fabric palette and create a true story. Prints, woven, or printed shirting stripes, multicolored embroideries, jersey knits with various colors and even beading layouts can all serve as an axis to bring unrelated colors together and harmonize the group.

5.2   Color science

5.2.1   Color theory

- Primary colors: colors that cannot be created by combining any other colors or hues.Red, yellow and blue;
- Secondary colors: colors that are created from a combination of two primary colors.Green, orange and purple;
- Tertiary colors: color that are formed by mixing a primary color and secondary color. Yellow-orange, red-orange, red-purple, blue-purple, blue-green, and yellow-green;
- Color wheel: a kind of circular arrangement of the spectrum, colors are arranged in the order in which the spectrum appears in nature;
- Complementary colors: two colors that are opposite each other on the wheel. Placing them next to each other makes for maximum vibrancy;
- Analogous colors: colors that are next to each other on a twelve-part color wheel (Figure 5.1,

Cover 2);

- Tints: colors mixed with white;
- Shades: colors mixed with black;
- Tone: a general term to describe the level of shade or tint;
- Hue: the particular gradation, or the variety of the color. Colors that have a hue are called chromatic colors. Black, white, and gray are neutral or achromatic, having no hue;
- Patina: the surface or texture of the color, often associated with the aging process;

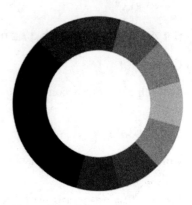

Figure 5.1　Color wheel

- Chroma: chroma refers to a color's saturation, or degree of departure from the neutral of the same value, the purity of a color in relation to gray;
- Saturation: the purity and density of a color.

5.2.2　Popular colors and their frequently associated meanings

- White: purity, surrender, truth, peace, innocence, simplicity, sterility, coldness, death, marriage (Western cultures), birth, virginity;
- Black: intelligence, rebellion, mystery, modernity, power, sophistication, formality, elegance, evil, death, slimming quality (fashion), occult;
- Gray: elegance, conservatism, respect, wisdom, old age, boredom, dullness, pollution, neutrality, formality, blasé, decay, military, education, strength;
- Red: passion, strength, energy, sex, love, romance, speed, danger, anger, revolution, wealth (China), marriage (India);
- Orange: happiness, energy, balance, heat, enthusiasm, playfulness, warning, fall, desire, optimism, Protestantism, abundance;
- Yellow: joy, happiness, summer, cowardice, illness, hazards, greed, femininity, friendship (During a time of war, yellow ribbons symbolize the hope for troops to return home);
- Green: nature, fertility, youth, inexperience, environment, wealth, generosity, jealousy, illness, greed, growth, health, stability, calming, new beginnings;
- Blue: water, oceans, peace, unity, calmness, coolness, confidence, conservatism, loyalty, dependability, idealism, depression, sadness;
- Purple: nobility, envy, spirituality, creativity, mystery, wisdom, gaudiness, exaggeration, confusion, pride, instability;
- Brown: nature, richness, rusticism, tradition, boorishness, dirt, dullness, filth, heaviness, poverty, roughness, earth, comfort (Figure 5.2, Cover 2).

Figure 5.2   Different colorways of popular colors

5.3   Color management

5.3.1   What is color management?

Color management is about how color trends evolve, the resources on which designers and merchandisers rely in order to select their brands' seasonal color palettes. Color management is the process of controlling the outcome of a color, from the initial concept (a chip, swatch, yarn, or sample) to the final production output, in a way that is acceptable to the consumer.

5.3.2   Color management considerations (ref to relating color palettes to target markets)

In today's marketplace, a well-defined color palette should take into consideration the target market's age and life stage, fashion level, coloring, ethnic diversity, geography and climate, and the garment's function.

Age and life stage

Target customers groups have unique preferences when they are considering color. Young adults are less constricted by dress codes and desire to change, so they often see clothing purchases in terms of a single-season life span rather than several years.

As consumers enter the full-time workforce or begin families, they enhance their personal coloring and the confidence to wear colors instead of blindly following fashion trends. They shop for clothes with the intention of wearing them for more than one season.

Fashion level

Product developers at each fashion level may interpret the color trends introduced by couture and designer markets. In the given market, there are customers who are more fashion forward and others who are more conservative. Therefore, a color palette should accommodate these preferences as many as possible.

Personal coloring

Given the skin tones and hair coloring of the market's customers are different, groups within a seasonal line should offer a balance of warm and cool colors to provide compatibility. Most colors can be adjusted through manipulation of value and chroma to look becoming with all complexions and hair colors.

Geographic location

Since the climates and regional preferences are different geographically, color stories should be carefully decided to fit the geographic markets appropriately. In a global market, seasonal considerations can extend the life cycle of a garment because seasons are reversed above and below the equator. For instance, in China and Japan, consumers associate gray with inexpensive products, while gray connotes quality in the U.S.; conversely, the color purple is considered cheap in the U.S. and expensive in Asian countries.

End-use

Each category of apparel (swimwear, active sportswear, dresses, outerwear) interprets color somewhat differently.

5.3.3 Managing the color story

As the seasonal color palette is broken down into deliveries and groups, each is developed around a specific theme. Each color is named to identify it throughout the process of ordering fabric and materials to selling. Some companies use generic color while some fashion forward companies name their colors to further develop their concept or theme. An orange red might be named flame or poppy, evoking a visual image. These color names are a marketing tool that can be used in the showroom and in writing copy for catalogs, websites, and print advertising.

The size (breadth or depth of selections) of the lines to be offered are typically decided early in the product development process by upper management or the merchandising staff. Multi-brand product developers may manage seasonal color palettes of up to 400 different colors. Each color may start with a paint chip, a piece of fabric, a hank of yarn, or a photo, of which the color impression can be translated into a standard that the entire supply chain can understand.

◎ Summary

Color decisions are based on both consumer demands and production efficiencies. Color that is worn becomes part of the wearer's physicality and personality. Color management is about how color trends evolve, and the resources on which designers and merchandisers rely in order to select their brands' seasonal color palettes. Color stories should be carefully decided to fit the geographic markets appropriately.

◎ Key terms

Achromatic 无色彩   Lab dips 色样

Additive color mixing system 加色混色系统   Pigment 色素

Chroma 色度   Primary colors 一次色

Color temperature 色温   Secondary (intermediate) colors 二次色

Color wheel 色轮   Shade 灰度

Complementary 互补的   Tint 色彩

Hue 色调   Value 色值

Illuminants 明度

◎ Product development team members

The following product development team members were introduced in this chapter:

Textile colorist（面料色彩师）: chooses the color combinations that will be used in creating designs.

Product developers（产品开发者）: people who are at each fashion level and interpret the color trends introduced by couture and designer markets.

◎ Activities

1. Select a color story from a color trend forecast resource at your school. Mix those colors on your computer screen and save them as a color palette.

2. Shop a store such as the Gap, the Limited, Old Navy, or American Eagle whose product consists of store brands. Identify the colors they have chosen for their seasonal color palette.

◎ Review or discussion questions

1. Discuss how the seasonal colors for the current seasons vary between discount, moderate, and better stores in your area.

2. What factors may cause materials that were matched to the same standard to appear as different colors?

◎ Semester project

Semester project II : Create a trend board—Part 2

1. Your color scheme mood board will come to life with color swatches, photos of anything that reflect your color inspiration including landscapes and products.

2. Add a few more touches to capture the mood of the color scheme you've envisioned.

3. Text tool to annotate your design. The text on a mood board is used sparingly but can be perfect for expressing a concept or tying all of your ideas together.

Example (Figure 5.3, Cover 2) :

Figure 5.3   A trend board

◎ References

[1] MACBETH G. Fundamentals of Color and Appearance[M]. New York: Gretag Macbeth, 1998.

[2] KADOLPH S. Quality Assurance for Textiles and Apparel[M]. New York: Fairchild Books, 1998.

[3] LONG J, LUKE J T. The New Munsell Student Color Set[M]. New York: Fairchild Books, 2001.

[4] METHA P. An introduction to quality control for the apparel industry[J]. Milwaukee:ASQC Quality Press, 1992.

[5] P J, PARK K. Pick a shade-any shade?[J]. International Dyer, 2005, 190(4): 32-34.

[6] AGARWAL N. A note on color inconstancy[J]. Colourage, 2004, L1(10): 3.

[7] AZOULAY J F. The devil in the details: The challenge of color[J]. AATCC Review, 2005, 5(2): 9-13.

[8] AZOULAY J F. Color, light, and getting it right[J]. AATCC Review, 2005, 5(4): 40, 42-45.

[9] WELLING H. Color blind? Rethink your color processes![J]. Apparel Industry Magazine, September, 1999.

◎ Useful websites

[1] Pantone. 2006. www.pantone.com.

[2] SCOTDIC.www.scotdic.co.uk/E06.html/.

[3] The Spruce. www.thespruce.com.

[4] A to Z Color Consulting. www.atozcolor.com/color.html. Introduction to color theory, monitor calibration and color management.

[5] Color Marketing Group, www.colormarketing.org. Web site for the Color Marketing Group.

[6] Color Matters,www.colormatters.com. Web site of J. L. Morton, professor of color and color consultant.

[7] Color Pro,www.colorpro.com. Good source for information on color technology.

[8] Color Vision. www.handprint.com/HP/WCL/wcolor.html. Useful information on color science and color theory.

[9] Color Association. www.colorassociation.com/site. Web site for The ColorAssociation of the United States (CAUS).

[10] Color Group, The (Great Britain). www.city.ac.uk/colourgroup/.The ColourGroup's Web site provides access to useful color diagrams and other links.

[11] Commission Internationale de l' Eclairage [International Commission on Illumination]. www.cie.co.at/cie/home.html. This Web site provides access to *CIE News*, the quarterly news bulletin, as well as to technical reports, standards, and conference proceedings.

[12] Concept2Consumer.www.aatcc.org/igroups/c2c.htm.Concept2Consumer is acommittee of the American Association of Textile Chemists and Colorists that is dedicated to the concerns of Product Development for textiles, apparel, and home furnishings.

[13] Datacolor.www.datacolor.com. Datacolor is a leader in intelligent color management

providing color matching software, on-screen color simulation software, shade sorting software, spectrophotometer calibration software, and more.

[14] EFG' s Color Reference Library. www.efg2.com/Lab/Library/index.html. Resource for books on color and a variety of color links.

[15] EWarna. www.ewarna.1a.com.

[16] Gretag Macbeth. www.grctagmacbeth.com. Information on Gretag Macbeth color management products and lasses.

[17] Inter-Society Color Council. www.1SCC.org. Access to online quiz on common color myths.

[18] Munsell.www.gretagmacbeth.com/index/products/products_colorstandards, html. Information on the Munsell Color System.

◎ Case study: Pantone color system

Pantone LLC is a limited liability company headquartered in Carlstadt, New Jersey. The company is best known for its Pantone Matching System (PMS), a proprietary color space used in a variety of industries notably graphic design, fashion design, product design, printing and manufacturing and supporting the management of color from design to production, in physical and digital formats, among coated and uncoated materials, cotton, polyester, nylon and plastics.

Pantone began in New Jersey in the 1950s as the commercial printing company of brothers Mervin and Jesse Levine, M & J Levine Advertising. In 1956, its founders, both advertising executives, hired recent Hofstra University graduate Lawrence Herbert as a part-time employee. Herbert used his chemistry knowledge to systematize and simplify the company' s stock of pigments and production of colored inks; by 1962, Herbert was running the ink and printing division at a profit, while the commercial-display division was US$50,000 in debt; he subsequently purchased the company' s technological assets from the Levine Brothers for US$90,000 (equivalent to $6,000,000 in 2019) and renamed them "Pantone".

The company' s primary products include the Pantone Guides, which consist of a large number of small (approximately 6 inches × 2 inches or 15 cm × 5 cm) thin cardboard sheets, printed on one side with a series of related color swatches and then bound into a small "fan deck". For instance, a particular "page" might contain a number of yellows of varying tints.

The idea behind the PMS is to allow designers to "color match" specific colors when a design enters production stage, regardless of the equipment used to produce the color. This system has been widely adopted by graphic designers and reproduction and printing houses. Pantone recommends that PMS Color Guides be purchased annually, as their inks become yellowish over time. Color variance also occurs within editions based on the paper stock used (coated, matte or uncoated), while interedition color variance occurs when there are changes to the specific paper stock used. The Pantone Color Matching System is largely a standardized color reproduction system. By standardizing the colors, different manufacturers in different locations can all refer to the Pantone system to make sure colors match without direct contact with one another (Figure 5.4, Cover 2).

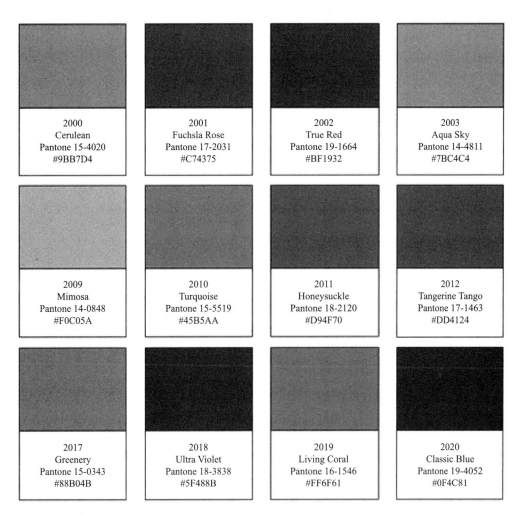

Figure 5.4   Pantone color system

One such use is standardizing colors in the CMYK process. The CMYK process is a method of printing color by using four inks—cyan, magenta, yellow, and black. A majority of the world's printed material is produced using the CMYK process, and there is a special subset of Pantone colors that can be reproduced using CMYK. Those that are possible to simulate through the CMYK process are labeled as such within the company's guides.

However, most of the Pantone system's 1,114 spot colors cannot be simulated with CMYK but with 13 base pigments (14 including black) mixed in specified amounts. The Pantone system also allows for many special colors to be produced, such as metallics and fluorescents. While most of the

Pantone system colors are beyond the printed CMYK gamut, it was only in 2001 that Pantone began providing translations of their existing system with screen-based colors. Screen-based colors use the RGB color model—red, green, blue—system to create various colors. The (discontinued) Goe system has RGB and LAB values with each color.

Pantone colors are described by their allocated number (typically referred to as, for example, "PMS 130"). PMS colors are almost always used in branding and have even found their way into government legislation and military standards (to describe the colors of flags and seals). In January 2003, the Scottish Parliament debated a petition (reference PE512) to refer to the blue in the Scottish flag as "Pantone 300". Countries such as Canada and South Korea and organizations such as the FIA have also chosen to refer to specific Pantone colors to use when producing flags.

Chapter 6

Fabrication

课题名称：Fabrication

课题内容：时装面料

课题时间：4 课时

教学目的：

1. 了解不同面料的分类。

2. 了解不同生产商挑选面料的过程。

3. 认识面料的美感、视觉效果以及功能性等特点。

4. 认识新技术对面料选择的影响。

5. 理解印花特点的重要性，包括如何挑选、设计和生产。

6. 了解面料设计中的版权法律。

7. 理解面料选择规划中的各项职责。

6.1 Creating a fabric story

6.1.1 Fabrication and fabric story

Fabrication is a process of the selection of textiles and trims for each style and grouping in a line.

Fabric story is an integral reference plan for fabric and trim selection, which includes concept, brand identity, current trends, customer's needs, color story, related materials, and other factors.

6.1.2 Why fabric story is important?

Learning how to put together a fabric story is an essential skill at the core of fashion design. A successful fabric story will:

- Support your concept and provide cohesion for the collection;
- Be consistent with the designer's identity and image;
- Address your customer's needs, aesthetics and lifestyle;
- Contain a diverse range of fabric weights and textures;
- Innovate fashion through application and/or manufacturing techniques;
- Address current trends while also providing an impetus for future ones.

6.2 Fabric and trim selection process

Figure 6.1 illustrates the fabric and trim selection process along with the line development.

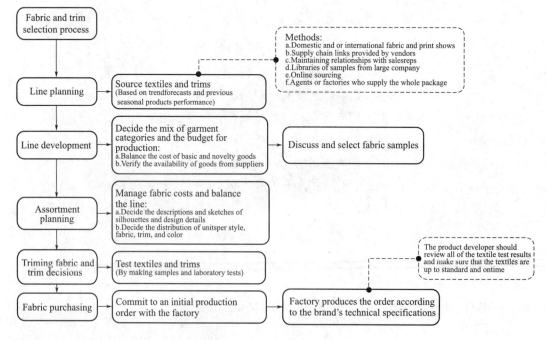

Figure 6.1 Fabric and trim selection process

6.3   Selecting textiles for their properties

6.3.1   Fabric properties concerned in fashion

Textile manufacturing has four stages: raw fiber production, yarn spinning, fabric construction, and finishing. A property can be introduced or removed, enhanced and diminished at each stage.

Fiber

Fiber is the raw material that makes up textiles. It is small and flexible enough to be spun into yarns and stable enough to be woven into fabric. There are many properties caused by the chemical and physical characteristics of the fibers.

Chemical structure. All fibers are made of long chains of molecules called polymers. The primary bonds that hold molecules together determine whether a fiber is strong, flexible, absorbent, and so on.

Physical structure. Fiber length is determined by source: natural fibers can be short or long; manufactured fibers can be infinitely long or cut into staple. Long filament fibers make smoother, stronger yarns; short stable fibers absorb more light and moisture and are more flexible. The size of fibers in very general terms is referred to as fine, medium, or coarse. The coarser the fiber, the stronger it is. The cross-sectional shape can influence the form of the surface that will lead to different properties.

Fibers are classified into two main categories: natural and synthetic. Natural fibers include cotton, wool, silk and flax. Synthetic fibers are chemically produced.

Yarn

Fibers are usually twisted together and spun into yarns. The inherent properties of the fiber can be added or diminished by the amount of twist: low twist makes yarns soft but weak; very high twist makes yarns bouncy and dull. About the methods of measuring yarn size, in general, the finer the yarn, the more flexible it is; the coarser the yarn, the stronger. There are two broad categories of yarns: spun and filament. Spun yarns usually are made of staples.

Construction

The characteristics of a textile can be affected by the arrangements of the yarns used in each type of construction. The three basic arrangements are woven, knit, and non-woven (Figure 6.2).

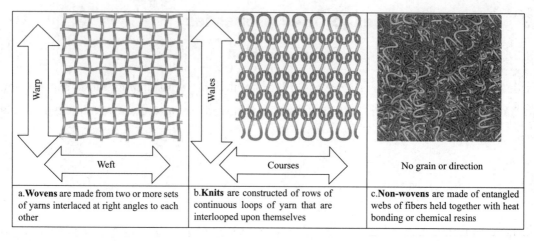

| a.**Wovens** are made from two or more sets of yarns interlaced at right angles to each other | b.**Knits** are constructed of rows of continuous loops of yarn that are interlooped upon themselves | c.**Non-wovens** are made of entangled webs of fibers held together with heat bonding or chemical resins |

Figure 6.2   Three basic arrangements

Coloring and finishing

There are two methods to add color to textiles: dyeing or printing. Dyeing can be accomplished at any stage during textile manufacturing: fiber, yarn, fabric, and even garment. Patterns can be applied to fabrics by various printing processes: screen printing, roller printing, and heat-transfer printing.

Finishes not only can increase or diminish the properties of textile, but also can add properties that are not inherent in the fiber, yarn, or construction. Common finishing methods that add or remove properties include chemical finishes, heated rollers, printing, brushing, and laminated surfaces.

6.3.2 Aesthetic properties

Figure 6.3   The fabric is a balanced plain weave so the yarns are the same in the warp and the weft. The flutes are evenly distributed

Aesthetic properties are expressed in apparel products through drape and hand, luster, surface interest, and other properties to convey meaning to the consumer.

Drape

The drape is the tendency of textile to cling to or stand away from a body when acted upon by gravity. We can call the drape is stiff, hard, firm, or crisp when a textile resists gravity. And when the textile gives in to gravity, we can call the drape is soft, clingy, swingy, bouncy, or limp. Hand (handle) is the property that is experienced when a textile is crushed in the hand and then released. In general, soft drape and hand are usually found together. The drape of a fabric can be the result of fiber, yarn, construction, finish, or support material. Figures 6.3 through 6.5 show various dress drape.

Figure 6.4   The drape of a fabric can change dramatically when the warp and weft are made of different weight yarns. These fabrics are all unbalanced plain weaves. The heavier yarns can hold out the fabric in wide or stiff folds and the finer yarns allow the fabric to hang in narrower flutes

Luster

Luster is an aesthetic property that is perceived as the amount of light reflected from the textile surface. A smooth surface can reflect a lot of light. Conversely, textiles with deep, open areas on the surface will absorb light. The luster of fabrics is influenced by the fiber, yarn, construction, or finishing (Figure 6.6, Cover 2).

Surface interest

Surface interest in textile is created by breaking up a surface into contrasting units of light reflection: shiny/dull, rough/smooth, high value/low value, high chroma/low chroma. When surface interest is introduced into a textile before the finishing stages, it is called structural design, otherwise it is called applied design. The factors influenced on the surface interest include fiber, yarn, construction, finish, and trim (Figures 6.7-6.10).

Figure 6.5   Floats are formed in a woven textile when a yarn does not interlace with the yarn running in the opposite direction for multiple yarn crossings. The floats move more freely and allow the fabric to drape along the grain in which they are woven

6.3.3   Utilitarian properties

Utilitarian properties, which include safety properties, comfort properties, and other properties, are characteristics that make the product useful to the consumer.

Figure 6.6   The luster of fabrics

Figure 6.7   Novelty yarns increase surface interest

Stretch

Stretch imparts the abilities to a yarn, fiber, or textile both to elongate with stress and to recover its original shape and size when the stress is removed. It is an important property that influences the comfort level of a garment. Stretch can vary for many reasons, such as the characteristics of the fiber, yarn, construction, and trim. For example, woven fabrics can get "comfort stretch" by using stretch yarns; some knit constructions allow the loops to open up when stressed.

Figure 6.8   Warp arrangements increase surface interest in wovens

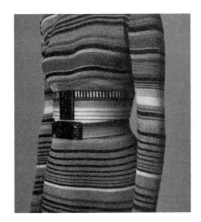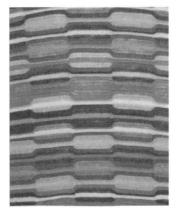

Figure 6.9   Complex patterns increase surface interest in knits

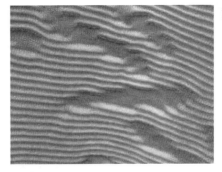

Figure 6.10   Applied finishes add surface interest

Durability

Durability is the strength of a textile exhibited through time and use. If a textile can resist the mechanical stress of abrasion, cutting, ripping, and tearing, it will be durable. Once the fiber is weakened or destroyed, textile can lose its durability. Resistance to various threats can be added to textile during each stage of production (Figure 6.11).

Figure 6.11　Textile construction can enhance resistance to abrasion, tears, and dirt

Weight

Weight is a subjective response to what the user senses. A down jacket may appear to be heavy because of its visual bulk, but it may feel surprisingly light when you wear it. For competitive athletes, weight has become their focus, because they want to reduce the weight that might impede their performance as much as possible. In general, fiber and trim are the main factors that may influence the weight of clothing.

Thermal management

Thermal management is a category of textile properties that give apparel products the ability to keep the body at its ideal temperature regardless of exterior conditions, activity level, or health. These properties include heat retention and conduction and are affected by moisture absorption (or repellency) and wicking. Common methods for enhancing thermal properties include fiber, construction, finish, and trim. For instance, weaving and knitting patterns can add tiny pockets that trap body heat to warm the wearer, and sometimes can also let body heat pass through (Figure 6.12).

Figure 6.12　Textile construction can enhance thermal properties

6.4   Common fabrics and trims

6.4.1   Common fabrics

The following are some common fabrics used in design.

Canvas

Canvas is a plain-weave fabric typically made out of heavy cotton yarn and, to a lesser extent, linen yarn. Canvas fabric is known for being durable, sturdy, and heavy-duty.

Cashmere

Cashmere is a type of wool fabric that is made from cashmere goats and pashmina goats. Cashmere is a natural fiber known for its extremely soft feel and great insulation. The fibers are very fine and delicate, feeling almost like a silk fabric to the touch.

Chenille

Chenille is the name for both the type of yarn and the fabric that makes the soft material. The threads are purposefully piled when creating the yarn, which resembles the fuzzy exterior of the caterpillar.

Chiffon

Chiffon is a lightweight, plain-woven fabric with a slight shine. Chiffon has small puckers that make the fabric a little rough to the touch. These puckers are created through the use of s-twist and z-twist crepe yarns, which are twisted counter-clockwise and clockwise respectively.

Cotton

Cotton is a staple fiber, which means it is composed of different, varying lengths of fibers. Cotton is made from the natural fibers of cotton plants. Cotton is primarily composed of cellulose, an insoluble organic compound crucial to plant structure, and is a soft and fluffy material.

Lace

Lace is a delicate fabric made from yarn or thread, characterized by open-weave designs and patterns created through a variety of different methods.

Merino wool

Merino wool is a type of wool gathered from the coats of Merino sheep. Merino wool is one of the softest forms of wool and doesn' t aggravate the skin. Merino wool is known for being odor-resistant, moisture-wicking, and breathable.

Polyester

Polyester is a man-made synthetic fiber created from petrochemicals. Polyester fabric is characterized by its durable nature; however it is not breathable and doesn't absorb liquids, like sweat, well.

6.4.2　Common trims

Trims are material components that are added on the garment. They are directly attached to the fabric to make garments. Trims can be threads, buttons, lining, beads, zippers, motifs, patches, etc. They add a style quotient to the overall look of the wearer (Figure 6.13, Cover 3).

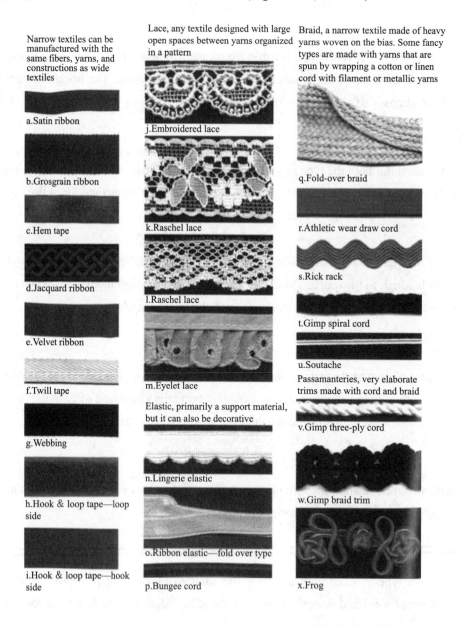

Narrow textiles can be manufactured with the same fibers, yarns, and constructions as wide textiles

a. Satin ribbon

b. Grosgrain ribbon

c. Hem tape

d. Jacquard ribbon

e. Velvet ribbon

f. Twill tape

g. Webbing

h. Hook & loop tape—loop side

i. Hook & loop tape—hook side

Lace, any textile designed with large open spaces between yarns organized in a pattern

j. Embroidered lace

k. Raschel lace

l. Raschel lace

m. Eyelet lace

Elastic, primarily a support material, but it can also be decorative

n. Lingerie elastic

o. Ribbon elastic—fold over type

p. Bungee cord

Braid, a narrow textile made of heavy yarns woven on the bias. Some fancy types are made with yarns that are spun by wrapping a cotton or linen cord with filament or metallic yarns

q. Fold-over braid

r. Athletic wear draw cord

s. Rick rack

t. Gimp spiral cord

u. Soutache

Passamanteries, very elaborate trims made with cord and braid

v. Gimp three-ply cord

w. Gimp braid trim

x. Frog

a.NATURAL MATERIALS. These materials can be used for buckles, cord ends, zipper tabs, etc.

| Shell | Horn | Leather | Wood | Tropical nut | Bone | Glass | Cloth covered |

b.METAL FINISHES. These finishes can be used for any metal hardware including buttons, buckles, rivets, snaps, zipper pulls, d-rings, etc. Each vendor has its own names for these finishes. Finishes are electroplated unless listed as "enamel". Other finishes names include brushed, etched, polished, brilliant, shiny, etc.

| Silver | Matt silver | Antique silver | Enamel painted | Nickel | Black nickel | Gold | Antique brass | Gun metal | Copper | Antique copper | Titanium |

c.METAL FABRICATION   d.PLASTIC. Plastic surface can be smooth, matte, etched, or molded. Colors are unlimited

| Stamped metal | Molded metal | Imitation shell | Imitation horn | Imitation leather | Imitation tortoise | Etched | Dyed |

e.BUTTON SHAPES. The silhouette of the button adds surface interest. Novelty buttons can be any shape possible. The method of attachment also contributes to the look. Two-hole and four-hole buttons add the texture of the thread for casual loods. Dome and ball buttons attached by a hole under the button (shank) so the top is smooth for a more formal look

| Rimmed | Flat | Dome | Ball | Toggle | Novelty | Novelty | Novelty |

f.WESTERN LOOK   g.DRESSES

| Tack button | Snap | Rivet | Buckle | | Studs | Hook and eye | Buckle |

h.OUTDOOR LOOK

| Side release buckle | Cam lock | Ladder lock | Grommet | Cord lock | D-ring |

Figure 6.13   Various trims

◎ Summary

The fabrication process is a critical step in apparel product development. Fabric contributes to a garment's aesthetics, function and serviceability. The fabrics chosen for a seasonal line speak to the designer through their fiber content, weight, surface interest, and drape. These fabric characteristics influence how the designer will use the fabrics in garment silhouettes.

◎ Key terms

Aesthetic properties 美学性能   Cotton 棉、棉纤维

Construction 构成   Drape 悬垂

Durability 耐用性

Dyeing 染色

Fabrication 面料构成

Fabric story 面料故事

Fiber 纤维

Flax 麻、麻纤维

Heat-transfer printing 转移印花

Knit 针织、针织物

Luster 光泽

Non-woven 非织造、非织物

Printing 印花

Roller printing 滚筒印花

Screen printing 筛网印花

Silk 真丝

Stretch 伸长、弹性

Thermal management 热管理

Trim 辅料

Utilitarian properties 实用性能

Wool 羊毛

Woven 机织、机织物

Yarn 纱线

◎ Product development team members

The following product development team members were introduced in this chapter:

Textile designer（面料设计师）: creates original designs for the fabrics used in all sorts of industries; they can be surface designers, knitters, weavers, or embroiderers.

Textile stylist（面料风格师）: the creative person who modifies existing textile goods: alters patterns or prints that have been successful on the retail floor to turn them into fresh, new products; and may develop color alternatives.

◎ Activities

1. Select an existing brand and analyze the fabrics used in the line. Develop a fabric story for an upcoming season. Present your fabric story in a board that includes a picture that suggests a theme and a swatch of each fabric with color chips to indicate color range.

2. If you have access to a computer program that will put a design into repeat, take a single motif and explore how varying the arrangement, repeat, and layout gives you many patterns. Print three of your favorite results and present them to your class.

◎ Review or discussion questions

1. What print themes have you observed in apparel stores this season? Is there any correlation between the prints being used in menswear, children's wear, and women's wear?

2. Why do product developers expect their sourcing partners to take ownership of the textile specified for apparel products?

◎ Semester project

Semester project Ⅱ : Create a trend board—Part 3

1. Pull together reference images. You can use Pinterest or Instagram to find images you are really into and save them. Put them together to make a pleasing group.

2. You could request samples from fabric shops and place the swatches to see if the color and the texture work well together.

Example (Figure 6.14):

Figure 6.14　A trend board

◎ References

[1] PATTY B, JANNETT R. Ready-to-wear Apparel Analysis[M]. 3rd Edition. New Jersey: Prentice-Hall, 2013.

[2] CORCORAN, CATE T. Industry heavyweight launch Internet sourcing service[J]. Women's Wear Daily, 2004.

[3] CORCORAN, CATE T, et al. If we can build it, will they wear it?[J]. Women's Wear Daily, 2004.

[4] Cotton Incorporated's Lifestyle Monitor. Knit wits: sweaters make smart style sense this season[J]. Women's Wear Daily, 2006a.

[5] Cotton Incorporated's Lifestyle Monitor. Performing arts[J]. Women's Wear Daily, 2006b.

[6] KADOLPH S J. Quality Assurance for Textiles and Apparel[M]. Fairchild Publications, 2007.

[7] KHATUA S. Printing. Presentation given to Kohl's product development team by MTL Laboratories, 2001.

[8] SKREBNESKI V L. The art of haute couture[J]. New York: Abbeville Press, Inc, 1995.

[9] TUCKER R. Stretching products to meet needs[J]. Women's Wear Daily, 2006, 19(119): 10.

[10] WATKINS P. Fibres & fabrics[J]. Textile View, 2005.

[11] DUNNE L E, ASHDOWN S P, SMYTH B. Expanding garment functionality through embedded electronic technology[J]. Journal of Textile and Apparel, Technology and Management 2005,4(3).

◎ Useful websites

[1] Nano-Tex. www.nano-tex.com.

[2] Patagonia. www.patagonia.com.

[3] MsterClass. www.masterclass.com/.

◎ Case study: Fabric dealing center (FDC) fabric

Beijing Meilun Technology Co., Ltd. is initiated by experts in the textile and apparel industry, and is committed to building an internal trade and cross-border B2B e-commerce trading platform for global textile and apparel fabrics relying on the Internet, Internet of Things, big data, and artificial intelligence technology (Figure 6.15). The FDC fabric library gathers the world's largest spot material samples to form an online and offline fabric display for buyers and sellers to query for free; the FDC fabric trading platform uses big data to predict fashion trends and greatly improves textile printing and dyeing factories. The efficiency of the supply chain effectively eliminates the inventory of fabrics and accessories in the textile industry and the clothing industry; the FDC fabric trading platform develops a transaction mechanism that ensures the integrity, safety and efficiency of buyers and sellers, which greatly reduces the cost of obtaining information for both sides of the fabric, and greatly improves the fabric. Efficient and convenient transactions promote industrial upgrading. The FDC fabric library project will change the traditional

Figure 6.15　The logo of FDC

trading and operating modes of the textile and clothing manufacturing industry, and build a new pattern, new model, and new future for the textile and clothing industry that meets the needs of the times, technology, and national development.

FDC fabric library is favored by industrial clusters, and has begun to set up FDC fabric books in important domestic textile and apparel cluster areas such as North, Shanghai, Guangdong, Zhejiang, Jiangsu, Fujian, and core areas of international fashion industry clusters such as New York, London and Moscow. The FDC fabric library project accurately addresses the needs of regional and special fabric industry cluster upgrades, and realizes a new operation model that conforms to the regional economic development plan and integrates traditional industries with Internet + Internet of Things + big data + artificial intelligence. Through the FDC fabric library project, the business concept of "buy the world and sell the world" is realized in line with the international market, and truly realize the transformation and upgrading of the intensive, high-efficiency, low-cost, high-output, high-profit emerging textile and apparel and special fabric industrial clusters.

Meilun Technology will continue to integrate new technologies, new models, new supplies, and new markets on a global scale, interpret the traditional textile and garment industry with new perspectives and thinking, promote industrial upgrading and reform, promote the vigorous development of the regional industrial economy, and promote the global textile and garment industry culture. Design, product, market, and technology continue to integrate and develop, contributing to the progress of the global textile and apparel industry.

Chapter 7

Garment Style Design

课题名称：Garment Style Design

课题内容：服装款式设计

课题时间：4课时

教学目的：

1. 探索记录设计思考的方法并尝试款式变化。

2. 了解如何将设计元素与服饰美学结合。

3. 了解男装、女装以及童装的分类。

4. 了解每类廓形的款式变化。

5. 了解设计细节如何增加服装造型的内涵。

6. 了解设计细节如何影响服装成本和生产安排。

7.1 Why garment style design is important?

The fashion dynamic is characterized by perpetual change. Nowadays, as society becomes more casual, fashion has become increasingly item-driven. In an item-driven environment, customers tend to buy more single garments than coordinated ensembles, so the product developers must understand how to manipulate seasonal silhouettes to suit the needs of their consumers.

Successful silhouettes must:

- Match the lifestyle and aesthetic preferences of the target customer;
- Provide value in terms of quality and fit;
- Be produced within a targeted price point;
- Be produced to arrive on the retail floor as scheduled.

7.2 Methods for developing design ideas

7.2.1 Garment style design process and methods

The methods product developers use to translate their design research to a collection of silhouettes vary. Common approaches include:

- Studying primary resources, such as historic or cultural artifacts;
- Buying actual garments for their silhouette, fit, fabric, or detailing;
- Collecting runway images from the Internet or print resources;
- Sketching design ideas observed while doing market research;
- Experimenting with interesting sample fabrics;
- Exploring the application of new technology.

7.2.2 Studying primary resources

Studying primary resource material frequently results in the most original design ideas. High-end designers may get new ideas from traveling to exotic locales or viewing costume exhibitions featuring the works of great artists or designers. Some museums will make arrangements to allow designers to study the work of a particular period or designer in their archives.

7.2.3 Studying purchased garments

Product developers generally have budgets for purchasing garments. Those who work derivatively have the largest budgets because their lines are developed directly from the samples they buy. Buying samples abroad, rather than from domestic competitors, enables product developers to better differentiate their lines. Branded product developers tend to have smaller budgets; they only buy samples when they want to reinterpret a particular fabric, fit, detail, or construction technique that they cannot capture through sketching or photographs. Very small product developers generally have no budget for purchasing garments and are left to interpret trends in other ways.

7.2.4　Printed sources and the Internet

Product developers use these sources to create concept boards, a collection of images, sketches, and swatches that express the design direction they are exploring for a particular group.

7.2.5　Sketches

Silhouette ideas are generally developed as croquis sketches in a sketchbook. "Croquis" is the French word for sketch. In contrast with the realistic figure, fashion croquis figures are elongated, drawn anywhere from nine to twelve heads high.

To ensure that the designers' focus is on the design of the garment, not the proportions of the figure, they may rely on an underdrawing or lay figure—a well-proportioned pose that can be slid under a page and used as a template to help control proportions and the location of garment details.

When using an underdrawing, the design idea is drawn first and then the visible figures are added. There is no need to draw in facial features or hair. Some designers sketch over an underdrawing without adding the body itself. These sketches are called floats. Some designers prefer to develop their ideas as flats, two-dimensional drawings of garments that represent how the garment looks when spread out on a flat surface, rather than how the garment appears on the body. Digital flats can be filled with color or digital representations of the fabric in which they will be manufactured.

7.3　Fashion design elements

7.3.1　Definitions

Garment design is the selection and interpretation of color, fabric, styling, and fit.

Design is the organization of design elements, using principles, to create products that are aesthetically pleasing to observer.

Design elements are the building blocks of design. These elements—line, color, texture, pattern, silhouette, and shape—are intrinsic to every product, including apparel.

7.3.2　Line

Line determines the silhouette of the garment and the shapes formed within the garment. The use of line can create optical illusions on the body. The optical illusions work best when the other design elements are used to enhance the impact.

7.3.3 Color

How color is utilized in an ensemble can create figure illusion and color harmonies affect how we perceive the figure. Understanding how color affects the figure can help to provide options when selecting a color range for a particular silhouette or when determining how to best use color harmonies that are in fashion.

7.3.4 Texture

Texture is the term used to describe the surface or hand of a fabric and can be attributed to a combination of the fabric's characteristics—fiber, yarn, construction, weight, and finishing. The texture of a fabric affects how we perceive color.

7.3.5 Pattern

Pattern can be constructed into the fabric through weaving, knitting, or felting; or they can be applied to the fabric through printing, embossing, and other specialty techniques such as dévoré or laser cutting.

7.3.6 Silhouette

The garment silhouette is the outer shape of a garment. It is not always possible to see a silhouette clearly when a garment is on a hanger.

7.3.7 Shape

The silhouette is frequently sectioned off into smaller shapes within the garment using seam lines, details, and garment edges. These shapes can:

- Add styling interest to the silhouette;
- Help to achieve fit;
- Allow for the combination of two or more fabrics;
- Allow the designer to create optical illusions and/or enhance body proportions;
- Add functionality (such as pockets) ;
- Create symmetrical or asymmetrical balance;
- Create equal or unequal proportion;
- Create rhythm within the ensemble.

7.4 Fashion design principles

7.4.1 Design principle

The design process revolves around determining how to combine the design elements we have just reviewed into a pleasing whole. All those decisions are guided by an understanding of design principles which includes the following aspects.

7.4.2   Proportion

Proportion is the relationship or scale of all of a garment or ensemble's parts to each other and to the body as a whole. A garment is divided into sections by horizontal lines. It was believed that ratios of 2 : 3, 3 : 5 and 5 : 8 were the most pleasing to the eye, according to the rule of the golden mean of the ancient Greeks.

7.4.3   Balance

Balance refers to the distribution of visual weight of objects, color, texture, and bulk in a garment, giving it a sense of stability or equilibrium, determined by dividing a silhouette vertically down to the middle. Asymmetric garment can be readily mixed and matched with other symmetrical garments in the wardrobe. An asymmetric garment must be carefully thought through during the patternmaking and cutting processes. Coordinating garments should be similarly balanced or neutral with no visible center point.

7.4.4   Focal point and emphasis

A garment's focal point or emphasis is the first place on the garment to which the eye is drawn, like a convergence of the lines, a combination of colors, or a detail. If several of the design are competing for viewer's attention, the garment may be overdesigned.

7.4.5   Rhythm

Rhythm is the natural movement of the eye through the created elements of a garment. It can be achieved by strong silhouette lines; through the use of color, line, or shape; or through the use of repetition, radiation, or gradation.

7.4.6   Harmony and unity

Harmony means that all of the design elements work together in a garment to produce a pleasing aesthetic appearance and to give a feeling of unity to the design. To achieve this goal requires successful placement of a focal point that suggests rhythm.

7.5   Garment variations by category

7.5.1   General principle

When the garment fits close to the body, shaping devices are necessary. Shaping devices are darts, seams, pleats, and gathers that help to mold the garment to the contours of the body. The use of fabrics made with fibers such as Lycra also helps to shape garments.

7.5.2 Garment classification

Within each garment classification, there are certain elements that vary from season to season according to fashion trends and the specific needs of various target markets (Figure 7.1).

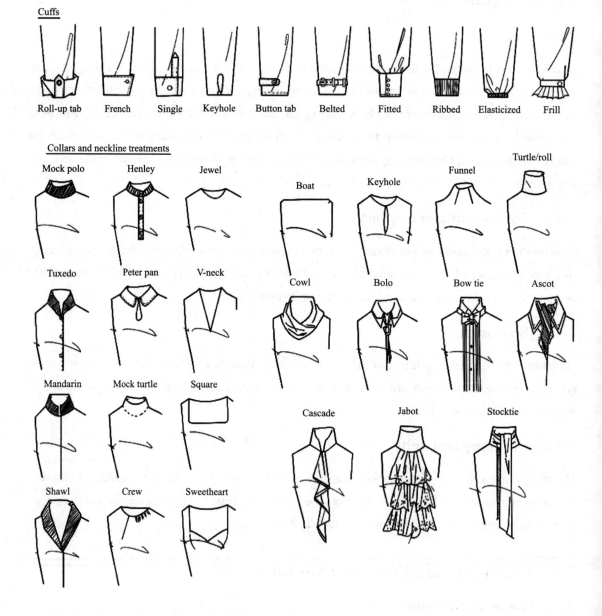

Cuffs

Roll-up tab French Single Keyhole Button tab Belted Fitted Ribbed Elasticized Frill

Collars and neckline treatments

Mock polo Henley Jewel Boat Keyhole Funnel Turtle/roll

Tuxedo Peter pan V-neck Cowl Bolo Bow tie Ascot

Mandarin Mock turtle Square Cascade Jabot Stocktie

Shawl Crew Sweetheart

Tops and details

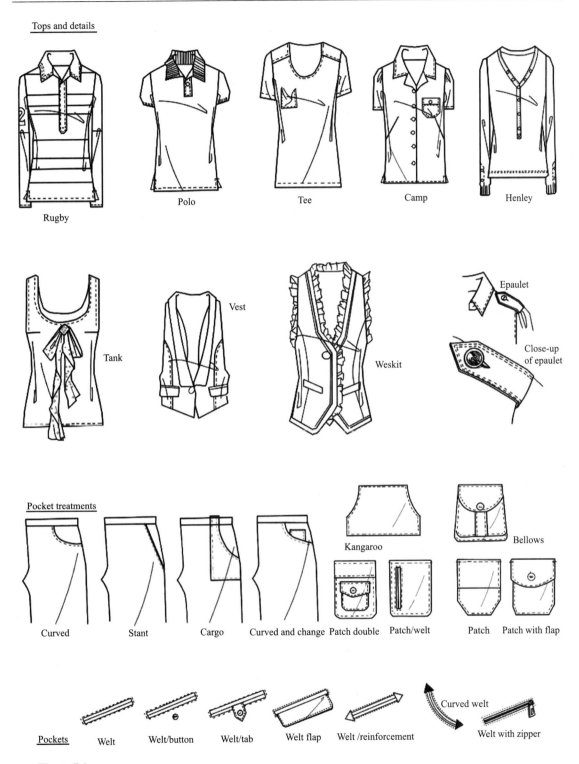

Rugby

Polo

Tee

Camp

Henley

Tank

Vest

Weskit

Epaulet

Close-up
of epaulet

Pocket treatments

Curved   Stant   Cargo   Curved and change   Patch double   Patch/welt   Patch   Patch with flap

Kangaroo

Bellows

Pockets   Welt   Welt/button   Welt/tab   Welt flap   Welt /reinforcement   Curved welt   Welt with zipper

Figure 7.1

Anatomy of a shirt

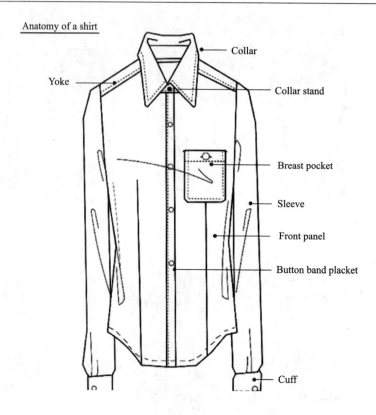

Collar

Yoke

Collar stand

Breast pocket

Sleeve

Front panel

Button band placket

Cuff

Blouses and shirts

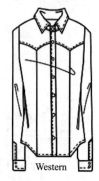

Shell tank

Western

Bowling

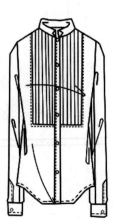

Sailor/midday

Peasant/gypsy

Poet/artist's smock

Tuxedo formal

Coats and outerwear

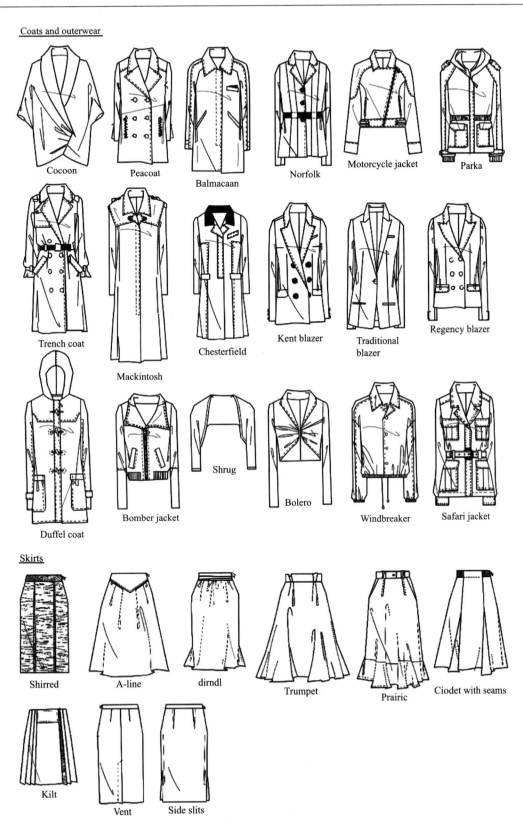

Figure 7.1

Box pleat   Side pleat   Knife pleat   Accordian pleats

Mushroom pleat   Flounce   Tiered

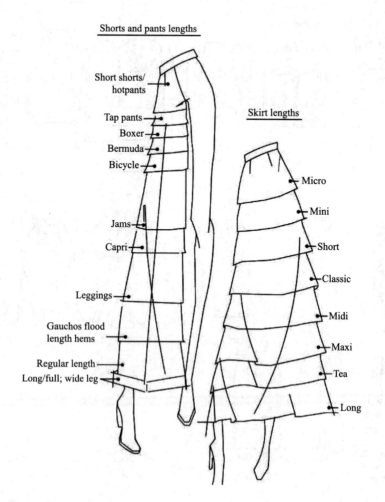

Shorts and pants lengths

Short shorts/hotpants

Tap pants
Boxer
Bermuda
Bicycle

Skirt lengths

Jams

Capri

Leggings

Gauchos flood length hems

Regular length
Long/full; wide leg

Micro

Mini

Short

Classic

Midi

Maxi

Tea

Long

Dresses

Halter

Wrap

Trapeze

Slip

Empire waist

Pouf

Sheath

Blouson

Cheongsam

Shirt dress

Halter

Menswear

Duffle coat

Parka

Trench coat

Mackintosh

Tuxedo jacket

Figure 7.1

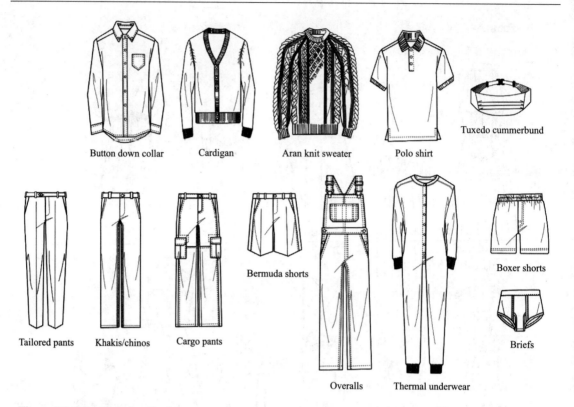

Figure 7.1   Garment classification

◎ Summary

The fashion dynamic is characterized by perpetual change. Garment design is a process that utilizes the design elements of line, color, texture, pattern, silhouette, and shape to create a garment. Designers explore garment ideas through the use of croquis sketches. The design process revolves around determining how to combine the design elements into a pleasing whole. The decisions are guided by an understanding of design principles. Garment silhouettes worn today are relatively standard, evolving slowly from season to season. Once a silhouette is determined, garments can be further embellished with details, including collar, sleeves, cuffs, pockets, trims, and closures. Understanding the style variables present in different garment classifications can help a designer be more creative.

◎ Key terms

Balance 平衡

Design 设计

Design elements 设计元素

Emphasis 强调

Focal point 焦点

Proportion 比例

Garment design 服装设计

Rhythm 节奏

Harmony 和谐

Shape 形状

Line 线条

Silhouette 廓形

Pattern 图案

Texture 材质

◎ Product development team members

The following product development team members were introduced in this chapter:

Product development manager（产品开发部经理）：the job of product development manager within the fashion/retail industry has a strategic overview of the whole company with the aim of moving the fashion brand forward by spotting new fashion trends and areas of growth which will be profitable for the fashion/retail company. The job role of a product development manager is to ensure the new product design is interpreted and manufactured to the correct standard.

Product developer（产品开发者）：a product developer will usually report into the product development manager. A design background, preferably within a fashion and retail environment is required for the job of product developer. The job role of the product developer within a fashion and retail environment includes producing detailed tech packs and following complete development of product from initial drawings to final sample, liaising with factories and also travel to offshore factories to follow progress and check production, sourcing trims and fabrics, ordering raw materials and communicating with vendors.

◎ Activities

1. Go out shopping for jackets. In your croquis book, sketch all of the variations you see being used to fit jacket bodies. Look for different combinations. Try on some of the jackets to evaluate the resulting fit. Take your tape measure to identify some key measurements for side-panel width and its relationship to two-piece sleeve seaming.

2. Select a trim, such as buttons, zippers, or ruffles. Experiment with using that trim as a design element in garments. Express your ideas in a series of croquis drawings.

3. Go shopping and study garments that utilize embroidery, applique, or beading. Design a garment that features one of these trims. Then develop the repeat of the trim and specify the stitches and components to be used.

4. Select a runway garment that you find in a magazine or on the Internet. Do a series of design sketches interpreting the runway design for a particular customer and price point. Then do a second series of sketches interpreting that same design for another market.

◎ Review or discussion questions

1. After reviewing fashion periodicals and shopping your local market, discuss the fashion direction that you see pants taking. In the women's market, are pants or skirts more important for the season you are studying?

2. What role do skirt lengths play in the current fashion season? Can you relate the prevalent length to fashion cycles or current events? Do you notice a difference in length offerings at different price points or for different markets?

3. Bring examples of garments with symmetrical and asymmetrical balance to class. Discuss your experience in wearing those garments. Do you wear some garments more often than others? Are some garments easier to wear than others? How long have you had each garment in your wardrobe?

4. Thinking about the current fashion season, is minimalism or embellishment more prevalent? Identify component parts, details, and trims that help to identify the season.

◎ Semester project

Semester project Ⅱ: Create a trend board—Part 4

Go to galleries, museums, fabric fairs, and shops. Listen to music, watch movies, see what's in style on the streets, look at architecture, design objects, industrial design, visit boutiques. Old photos, accessories, trims, basically any visual reference. Be open to what catches your interest and what you get ideas from.

Enjoy, this is one of the best things about being an apparel designer and entrepreneur. You are the one putting out the vision and the direction.

◎ References

1. KEISEE S, VANDERMAR D, GARNER M B. Beyond Design, the Synergy of Apparel Product Development[M]. 4th Edition. New York: Fairchild Books, 2017.

2. FAERM S. Fashion Design Course, Principles, Practice, and Techniques: The Practical Guide for Aspiring Fashion Designers[M]. 2nd Edition. Barron's Educational Series, 2017.

◎ Useful websites

[1] Online version of Elle.www.elle.com.

[2] An online fashion newsletter.www.fashion.about.com.

[3] An online fashion forecasting and trend reporting resource.www.fashiontrendsetter.com.

[4] Online version of Glamour.www.glamourmagazine.com.

[5] Online version of Instyle.www.instyle.com.

[6] Fashion&Retail. www.fashionpersonnel.co.uk.

Chapter 8

Line Development

课题名称：Line Development

课题内容：产品系列开发

课题时间：4课时

教学目的：

1. 了解产品预算和潮流趋势对生产链的影响。

2. 理解生产链不同阶段中的设计语言。

3. 理解单品/系列生产链的不同策略。

4. 了解服装和纺织品设计的法律保护范围。

5. 了解时尚发布。

8.1 Merchandise planning tools

Designing a group of garments that can be merchandised together is quite different from designing a successful garment. It requires designers and merchandisers to work as a team in order to coordinate and execute the merchandise budget, line plan, fashion forecast, assortment plan, and time and action calendar into product for a selling season.

Merchandise planning is the process of maximizing returns and minimizing losses on investment by planning assortments and production. Nowadays, the responsibility for merchandising usually lies with a team that includes the merchandiser, trend forecaster, designer, technical designer, and merchant/buyer. They work together to make the best possible decisions for the brand.

Merchandising involves:

- Evaluating historical sales and changes in sales potential;
- Summarizing information about the current business climate;
- Analyzing the trend forecast and making a sales forecast;
- Assessing the stock levels required to support the marketing strategy;
- Translating sales and profit goals into a line plan;
- Regular meetings with design to see how the line plan will be interpreted into product;
- Quantifying the seasonal line.

8.1.1 Sales forecast and merchandise budget

According to the profit goals, the company will develop a sales forecast and merchandise budget. **Sales forecast** is the prediction of achievable sales revenues based on historical sales date and market analysis. And merchandise managers will use the sales forecast to develop a merchandise budget to determine how to allocate resources to achieve the required inventory levels to meet sales goals.

8.1.2 The line plan

The sales forecast and merchandise budget are used to develop a line plan that identifies product parameters for the season. It will guide the design team to balance each seasonal delivery and clarify things such as:

- Number of deliveries;
- Range of fabrics required;
- Product assortments;
- Target cost and margins;
- Creative goals based on current season trends.

Each garment designed is part of a group that will be merchandised together on multiple channels. Merchandise in-store must carefully consider the choice of geographical location. Online assortments can be a good way to test new concepts and brand extensions, and learn more about customer preferences. Catalogs may be aimed at customers not likely to shop online.

8.1.3   Concept boards

The design team will take over the planning process and start by creating concept boards that illustrate the theme for the line. A concept board should convey the key colors, fabrics, silhouettes, and details of the theme that the designer has identified for each group in the line. It can be created on the computer or collaged using cut-and-paste methods that combine color chips, fabric samples, and trend research in form of tear sheets, digital images, or sketches from shopping trips. The concept board helps the designers focus their ideas and launches the design process (Figure 8.1).

Figure 8.1   Concept board

8.1.4　The assortment plan

Designers use the line plan as a template and concept boards as a creative guide to develop silhouettes. An **assortment plan** combines the parameters of the line plan with the concept boards as a unifying theme (Figure 8.2). It is a detailed visual tool that:

- Reflects color, fabric, and silhouette decisions for the brand in relation to current trends;
- Establish the right variety and balance for the brand's target customer;
- Serves as a checkpoint to ensure sufficient inventory will be available to support the merchandising plan and meet sales and margin goals;
- Anticipates sufficient assortment to meet the needs of multi-channel distribution.

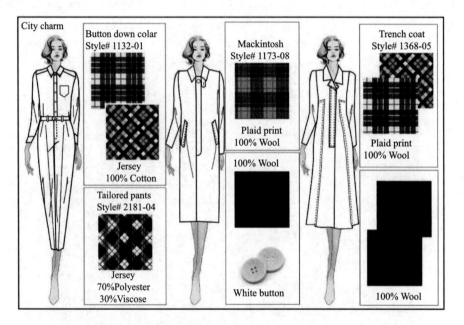

Figure 8.2　An assortment plan for a capsule collection

One traditional model for assortment planning is the pyramid plan. The lowest and broadest level of the pyramid consists of basic items (core items), which are usually low-risk and can be ordered in large quantities and assortments. In the middle level of the pyramid, product developers interpret current trends for their core customers, and these key items are critical to maintaining brand loyalty. The top level of the pyramid is made of fashion-forward or high-tech items that are not produced in great quantities. These items help to build brand image and enable customers to have access to new trends in fashion and technology.

However, increasingly, fashion-oriented brands prefer a diamond plan, in which the basic items only make up a small part of their product mix, key items make up the bulk of their products, and fashion-forward items are still represented at the top (Figure 8.3).

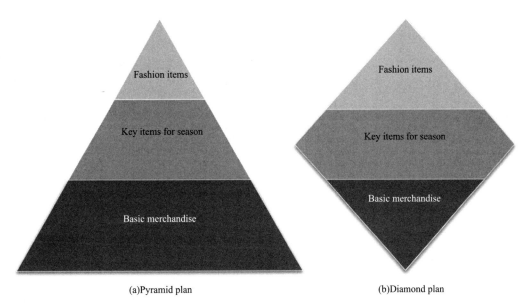

Figure 8.3　Pyramid plan and diamond plan

8.1.5　Methods of design development

In general, the design team develops newstyle groups or silhouettes for a product line using one of three methods or a combination of these methods.

Original designs

The first method is to create original styles through sketching and draping. Sketching designs using either hand or computer drawing techniques is the most common method in the industry today. Draping is usually used for customized products with higher price or for garments that reflect more fluid silhouettes.

Style modification

After modifications to the details of styling, color, or fabrication, styles from previous collections that sold well may be added to the new line. This method is very cost-efficient because it greatly reduces the overall development time, personnel efforts, and design costs such as pattern-making and fit-testing (Figure 8.4, Cover 3).

Knockoffs

Knockoffs are garments that are adapted or modified from products designed by other brands. They may be based on pictures of products or made from copying an actual product that has been purchased. Fast-fashion brands with flexible supply chains can interpret the ideas from designer runways for their own stores in six to eight weeks.

8.1.6 Concept review

Once the design idea is conceived, it must be sampled. For the design team, the **concept review** is a checkpoint at which they present their ideas to merchant buyers to ensure that design assortments are in sync with line plan expectations. Fabric, graphic, color, and silhouettes direction are approved at this time. Merchandisers, designers, and merchant buyers try to identify any voids in the collection, production challenges, or details that need to be refined. Once the concept review takes place, target prices are then determined based on preliminary costing discussions. The creative design team makes any agreed-upon adjustment to concepts, while the technical design team develops

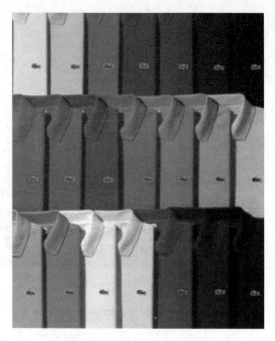

Figure 8.4 Polo shirts change little from season to season so can be offered in multiple colors

a spec package for the products that will move forward to the line review.

8.1.7 Line review

At this stage of process, the products are fully defined and costed through a technical package. It means that these products will be put into production. In line review, merchants and buyers quantify their orders by considering product needed for marketing campaigns, visual merchandising displays, and assortments required for their distribution plan. Once the line is quantified, sourcing will be responsible for production management.

8.1.8 Time and action calendar

The time and action calendar manages and assigns responsibilities in product development tasks to ensure delivery on the promised dates. It starts from market dates that are determined by the industry or, for retail product developers, the delivery dates when new merchandise must hit the stores. Since these dates coincide with promotional calendars, they are firm and non-negotiable.

8.2 Organizing the line

8.2.1 Item line development

Item line views each product as an item to be sold separately at a time. Dress, suits, coats, jeans,

T-shirts, and swimwear are common examples of categories developed as item lines. Item manufactures that specialize in a particular classification of product can take advantage of economies by specialized construction expertise and ordering large quantities of fabric at the best price.

In general, item lines have a limited range of product styles, but develop a wide range of color options. The coordinated color palette of various categories of goods enables them to be appealingly merchandised in a brand's stores, catalogs, and websites. These assortments have a longer shelf life, since their competitive edge is not fashion, but fit, proportions, styling, color assortment and quality (Figure 8.5, Cover 3).

Item lines are especially appropriate for juniors with limited disposable income and high interest in fashion. Product developers may adapt street fashion at the lowest price possible to satisfy customers' taste changes from season to season.

Figure 8.5　Classic polo shirts in different colors

8.2.2　Coordinated group line development

Coordinated group line is structured as groups of coordinating products and designed to be purchased as multiples. It usually consists of 30 to 50 pieces or more, including skirts, pants, jackets, tops, blouses, sweaters, and so on. The mix and match fabrics used in a coordinated group line allow customers to select different garments that complement their personalized appearance needs (Figure 8.6, Cover 3).

In general, new coordinated groups are delivered every 4 to 6 weeks, while fast-fashion suppliers introduce new items every two weeks. These small collections delivered more frequently are called capsule collections. To encourage customers to make purchases at full price when their size and styling preferences are still available,

Figure 8.6　Garments in a coordinated group line

items within the group will not be replenished and the remaining pieces will be marked down when new items arrive.

Coordinated groups are typically designed around a theme and expressed through colors, fabrics, stylings, or details. The advantage of this approach is that it helps to attract customers, and if one style in the group does not sell well, the style can be abandoned and the material assigned to it can be used to produce another style that sells better than expected.

8.3　Trends in the line planning environment

8.3.1　The evolution of pricing

A brand's price point is part of its identity. To a certain extent, it determines where the product will be sold, which brands to compete with, the range of fabrics that can be used, and who can afford it. Customers often believe that price is also a symbol of product quality.

In the past, designers who had runway shows might launch diffusion lines based on their signature lines to attract young customers. However, in recent years, new technology developments are impacting brands to develop the multi-channel platforms that customers now expect. Today, many luxury brands are combining their diffusion lines with their signature lines as they recognize that today's customer like to spend a lot of money on a famous brand for a special bag or a pair of shoes and then mix and match them with affordable basics. This phenomenon is often referred to as high-low dressing.

8.3.2　Co-branding

Entering into co-branding or exclusive brand agreements is another way for luxury brands to expand into new markets. Since these agreements are typically for a limited time, they require less infrastructure on the brand and they tend not to negatively impact the equity of the luxury brand.

8.4　Fashion presentations

Presentations during fashion week means huge expenditures. Luxury brands may host shows for both their signature lines and diffusion lines while contemporary designers are making themselves stand out by offering unique collections that are more in line with the needs of today's customers. If luxury brands want to maintain their influence, they must carefully consider and answer the following questions:

- Who are the shows for?
- Has the drama of fashion presentations overshadowed the clothes show?

- Are runway fashions reverent to the lifestyle of today's customer?
- Is it cost effective to host separate shows for a brand's signature line, diffusion line, women's wear, and menswear?
- Does the timing of fashion presentations make business sense?

◎ Summary

Merchandising a line begins with the sales and profit goals identified in the strategic plan. These goals are fleshed out in a line plan developed by the merchandising team in tandem with anticipated marketing promotions. Concept development takes place in a design team. A concept review and line review serve to edit these ideas into those products most likely to meet the expectations of customers and meet sales and profit goals. The line is organized as an item line or a coordinated group line.

The United states provides minimal legal protection for the design of apparel; somewhat more protection is granted to textile designs. The value of the significant investment required to produce a fashion show and the timing of those shows is being questioned.

◎ Key terms

Assortment plan 产品组合

Concept board 概念板

Co-branding 联合品牌

Coordinated group line 配套系列

Diamond plan 钻石方案

Fashion presentations 时尚发布

Item line 单品系列

Knockoffs 仿冒品

Line plan 系列计划

Merchandise planning 商品策划

Merchandise budget 商品预算

Original design 设计初稿

Pyramid plan 金字塔方案

Sales forecast 销售预测

Seasonal line 季节性系列

Style modification 样式修改

Time and action calendar 日程表

◎ Product development team members

The following product development team members were introduced in this chapter:

Creative designer (创意设计师): is responsible for carrying out the design direction in their department through original ideas for products.

Design director (设计总监): is an individual who guides the design team in developing creative concepts, making the lines developed by the company's various divisions more cohesive with regard to color, silhouette and fabric direction, adhering to the T&A calendar.

Director of product development (产品开发部主管): is responsible for the strategic planning of a product development division of a retail company.

Merchandisers (营销师): collaborates with the director of product development in deciding what to produce and then organizes and manages the entire product development process. Their responsibilities also include analyzing profitable and unprofitable designs and ensuring that the product development associates adhere to the T&A calendar.

◎ Activities

Visit an area mall. Shop a store that focuses on item lines. Identify several styles and how they are assorted by size and color. Then shop a store or department where the focus is on coordinated group lines. Study one group and identify the number of styles, fabrics, colors, and sizes in which this line is available.

◎ Review or discussion questions

1. Identify some of your favorite brands. Are they organized as item lines or coordinated group lines?

2. Discuss the variety required for a group line targeted to young adults just entering the job market versus a group line targeted to professionals who are well established in their careers.

3. What are some of the reasons for eliminating potential garment styles from a line, other than their aesthetic value?

4. Discuss the significance of fashion presentations as they now exist. Is the investment that a runway show requires worthwhile?

◎ Semester project

Semester project Ⅲ: Develop a line

Suppose you are going to develop a fashion brand, how would you develop a line? Here are some suggestions:

1. Write a business plan. This plan will act as a roadmap outlining how you'll reach your goals over the next couple of years.

2. Find your niche. Identify a problem within the market and then design a product expressly to fix that problem.

3. Understand your market. Study the companies whose product, marketing, and branding

strategies you admire, and whose target demographics you share.

4. Design and source the clothes for your line. Design your clothing and source your material to impress your consumer as the best quality product possible.

5. Price your products. Strike a balance between making a profit and setting a price that customers are willing to pay.

◎ References

[1] KEISER S, VANDERMAR D, GARNER M B. Beyond Design, The Synergy of Apparel Product Development[M]. 4th Edition. New York: Fairchild Books, 2017.

[2] MELANIE D, RIEGELMAN N. If Informed Fashion: Illustrations and Flats[M]. Los Angeles, CA: Denny & Riegelman Publications, 2003.

[3] JONES S J. Fashion Design[M]. New York: Watson-Guptill Publications, 2002.

[4] SORGER R, UDALE J. The Fundamentals of Fashion Design[M]. Switzerland: AVA Publishing SA, 2006.

[5] TAIN L. Portfolio Prentation for Fashion Designers[M]. 2nd edition. New York: Fairchild, 2003.

◎ Case study

Dior and I (Figure 8.7) is a 2014 documentary film written and directed by Frédéric Tcheng about designer Raf Simons' creative work for Christian Dior S.A. It premiered at the Tribeca Film Festival on 17 April 2014 and had a wide release in the United States on 10 April 2015. The film is centered upon Simons' debut season at Dior and includes non-speaking cameo appearances by Marion Cotillard, Isabelle Huppert, Jennifer Lawrence, Sharon Stone and Harvey Weinstein. The documentary received positive reviews by critics.

In this movie, you will see how a line sheet being formed from a trend board. Through it, you will experience the elegant and captivating part of the fashion industry. Of course it could help you understand this chapter better.

Figure 8.7 *Dior and I*

Part III

Technical and Production Planning

Chapter 9

Translating Concept to Product

课题名称： Translating Concept to Product

课题内容： 服装样板的概念

课题时间： 4课时

教学目的：

1. 了解服装样板的基本概念。

2. 从工业实际操作层面，阐述服装制板的方法和种类。

3. 了解服装实际生产过程中，服装合体性和质量方面的具体要求。

课前准备： 通过检索服装制作的视频，了解将设计图转换为工业样板的流程，建立初步的感性认识。

9.1   Patternmaking

9.1.1   What is pattern and patternmaking?

A **Pattern** is the template from which the parts of a garment are traced onto fabric before being cut out and assembled. The process of making or cutting patterns is sometimes called the one-word **patternmaking**, but it can also be written pattern making or pattern cutting.

A sloper pattern (home sewing) or block pattern (industrial production) is a custom-fitted, basic pattern, and many different styles can be developed based on this block pattern. Commercial clothing manufacturers usually employ at least one specialized patternmaker to make their own patterns as part of their design and production process. For customized clothing, slopers and patterns must be developed for each client, while for commercial production, patterns will be made to fit several standard body sizes.

9.1.2   The basic of patternmaking

Patternmaking is an art of manipulating and shaping a flat piece of fabric to conform to three-dimensional human figure. Essentially, patterns bridge the gap between ideation and production, which make this process straight forward for both the brand and the manufacturer. Patternmaking has become necessary for a fashion designer to enable him/her to convert a sketch into a garment via a pattern which interprets the design in the form of garment components, as shown in Figure 9.1.

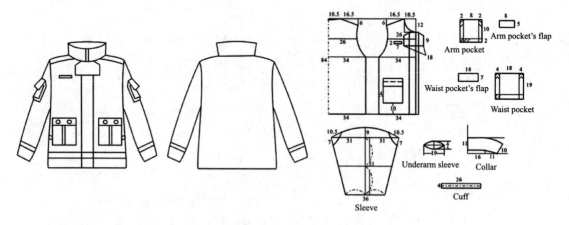

Figure 9.1   Translating design to pattern

9.1.3   Methods of creating patterns for first sample

Patterns for new, original garments may be created using drafting or draping methods. Patterns can be derived from existing sources using flat pattern, knock-off, or trace-off methods.

Pattern drafting

Pattern drafting is a process of translating a three-dimensional form into two-dimensional pattern. It involves measurements derived from sizing systems or accurate measurements taken on a person, mannequin, or three-dimensional scan. Measurements for chest, waist, hip and so on, and ease allowances are marked on paper and construction lines are drawn to complete the pattern. Drafting is used to create basic, foundation or design patterns.

Draping

Draping involves the draping of a two-dimensional piece of fabric around a form, conforming to its shape, creating a three-dimensional fabric pattern. Inexpensive substitute fabric such as muslin can be used for the original prototype. The advantage of draping is that the designer can see the overall design effect of the finished garment on the body form before the garment piece is cut and sewn (Figure 9.2).

Figure 9.2   Pattern drafting and draping

Flat pattern method

Flat pattern method involves the development of a fitted basic pattern with comfort ease to fit a person or body form. A block pattern (sloper) is the starting point for flat pattern designing. It is a simple pattern that fits the body with just enough ease for movement and comfort. The flat patternmaking method is widely used in the ready-to-wear market because it is fast, accurate and cost-effective. The important thing is to develop blocks that reflect the fit of target customers (Figure 9.3).

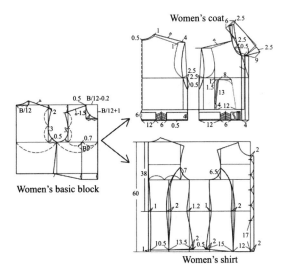

Figure 9.3   Converting block pattern into different garment styles

Knock-off

A knock-off is a direct copy of an existing style. It can be derived from a vintage garment, a competitor's product, or a style purchased from a brand that a company tries to simulate. The starting point can be a drafted pattern, a draped pattern, or modification to existing blocks, then the technical designer has to adjust the measurements to fit the firm's standard sample size, so the patternmaker can make a new pattern.

Trace-off

Another method for developing a pattern from one garment is called trace-off (rub-off or reverse engineering). It involves marking landmark points at various points on an existing garment and measuring the distance between these along lines that run horizontal and vertical to the center front of the garment. This method can be cost-effective for the production of basic garments, as it saves the development of the individual patterns or manipulation of blocks, and is often adopted by companies that do not have access to professional patternmakers.

9.1.4 Patterns for production

Once you have the initial basic pattern for your garments, the next step is to create alternative sizing for production. This process is called pattern grading. Some manufacturers have standard guidelines for grading, and this is where you see common retailer measurements. For example, sizes S, M, L, and so on.

Traditional methods

Manual method. After the first sample is made, reviewed in a fit session, and adjusted, and after a line presentation sample has then been made and adjusted, a final pattern is made in production fabric for approval before production starts. It will have accurate seam allowances, grain lines, notches, and all identifying information for each pattern part. A list of patterns called a pattern chart (cutter's must) is included. Once a production sample is made in the sample size, a complete set of production patterns is produced in a very labor-intensive process called grading. Each piece of the patterns has to be graded, labeled, and cut out in tag board.

Computerization. Pattern making using computer-aided design (CAD) software will become an easy job. Most companies use some kind of CAD software to create first patterns, production patterns, graded patterns, and markers. Companies can store both their basic blocks and past-season patterns so that they can be pulled up and modified to create new styles easily. Computerized systems have made the process of pattern making more economical and less time-consuming by improving fabric utilization and consistency. Even small start-up companies that cannot afford the hardware and software can contract specialty services that will digitize their handmade patterns so that the grading and marker making can be done electronically. Many computerized patternmaking systems are available in the market, including Gerber, Lectra, PAD, ET, Bock (博克), Richpeace (富怡).

New patternmaking technologies

Virtual three-dimensional draping. The cutting edge in pattern development is the evolving technology of draping styles directly on the computer. A garment may be designed through virtual three-dimensional modeling directly on the screen and the resulting style sketch literally unwrapped from the computerized figure and laid flat to form the two-dimensional outline printout for pattern pieces. This can be simulated using a 3D CAD system (Figure 9.4).

Three-dimensional patternmaking technology. Body scan technology is the most promising new patternmaking tool. It enables retrieval of three-dimensional body measurements from an individual consumer. These measurements may be directly applied to pattern development in flat-pattern, using hand or computerized techniques, to produce a customized product.

Figure 9.4   Virtual three-dimensional draping using Lectra

9.2   Standards and specifications

9.2.1   Standards

Standards provide parameters for making decisions about sizing, fit, and materials that reflect the needs of the target market while allowing the firm to make a profit. Some standards are developed over time as the company seeks its place in the market and always faces competition. Other standards are affected by governmental agencies where the product is produced or sold. Design specs are written by the designer or the technical designer in the context of these standards established by the company's strategic plan, including target customer, product category, distribution mix, price point, and branding strategy.

9.2.2   Specifications

Specifications should be prepared by product developers to make the sourcing partners understand the design concept. A combination of graphic and written instructions should be used. Specifications are not only for communication within the company, they also provide the basis for contracts with overseas vendors for materials, production capacity, quality standards, and evaluating vendor compliance. Every company will have a different name for the design spec and it will contain various levels of information, but at a minimum each style should have a style number, and include a technical flat front and back drawing, instructions about sewing and design details, a list of materials, a preliminary cost estimate, and a set of target measurements from which to create a first sample.

Technical flats

Technical flats visually define the proportions, details, and construction techniques required for production purposes as shown in Figure 9.5. They may include measurements and sewing notations. Complex details are sometimes enlarged in callouts (blowups), separate drawings in a larger scale that magnify an area so that a patternmaker or sample maker can understand exactly what is expected.

Materials

A design spec will list the textiles and trims needed to make the firsts ample. At a minimum, the design spec should list as much technical description of the material as possible or a source and order number of a product already identified from a textile show or sales rep.

Measurement chart

The design spec will include a measurement chart, a list of target finished garment measurements that guide the patternmaker to create a first pattern. Examples of key points to include in measurements

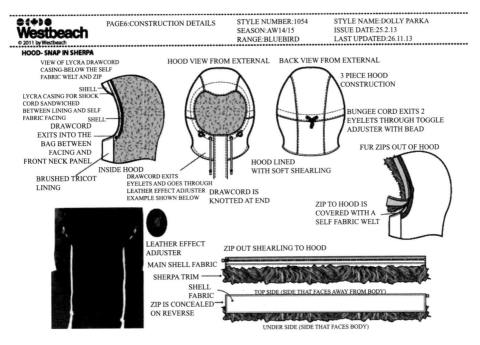

Figure 9.5   Tech packsheet

(Source: https://bentley-design.co.uk/tech-pack-example)

of shirts are: collar circumference or minimum neck stretch (in knits); bust or chest circumference; armhole, sleeve opening, and sleeve length; and center back length. Key points to measure on garment bottoms include waist, hip, front and back rise, functional zipper opening, and pant inseam or center back skirt length.

Cost estimate

Wholesale price is the amount charged by a wholesale product developer or manufacturer when the garment is sold to a retailer or distributor. These estimates are useful for determining whether a garment is suitable for remaining within the line and for comparing bids from competing vendors. Negotiating cost is often handled by production planners or sourcing specialists.

9.3 Sizing, fit and quality

9.3.1 Consumer perception of fit

More consistent and accurate fit and sizing systems are helpful for companies to compete and differentiate themselves in the market. Advances in technology can better meet fit requirements of consumers, however it will take strategic marketing plan to educate consumers about the new fit paradigm.

The concept of fit can be understood by issue that customers often describe.

- Consistency: garment does not fit like the same size in the same brand and another brand;
- Proportion: garment fits in one area of the body, but not all;
- Restriction: garment does not allow ease of movement;
- Fabric folds or bunches where it shouldn' t;
- Garment shifts to the front, back, or sides;
- Garment is not available in my size;
- Garment is too long or too short in body, legs, or arms.

These issues generally have easily recognizable causes that can be corrected by one or a combination of the following solutions.

- Establish the standard body size as a reflection of the target market, including body dimensions, shape, and proportions;
- Establish grade rules that reflect the target market;
- Refine patternmaking and construction techniques;
- Monitor compliance;
- Expand marketing strategies.

9.3.2 Standard body size

Once a brand has identified its target market, it will define that range of customers through body measurements. The brand will create a standard body size, which most closely represents the ideal shape of the target customer. Manufacturers will seek to minimize the number of sizes to keep inventories down, yet satisfy as many customers as possible. This list of sizes with corresponding body measurements is called a size system.

Fit model

Once the most common size in the size range is chosen, an effort will be made to determine the ideal shape, body proportions, and stature that represent that size. Companies may choose a fit model with the target girth and height measurements who projects an aspirational ideal that will appeal to the target market.

Dress form

Some companies use dress forms (mannequins) that represent the whole body or partial sections (such as torso or legs), which are placed on a stand to facilitate patternmaking, draping, and fitting activities. Forms are available for all genders and ages. Specialized forms are made for bathing suits, pants, maternity wear, and even dogs! For example, Alvanon developed serial dress forms as shown in Figure 9.6.

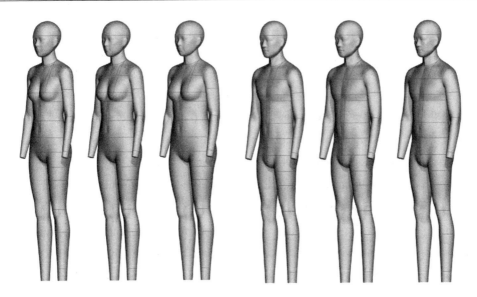

Figure 9.6   Serial dress forms of Alvanon

Three–dimensional body scans

Measurements are extracted from the three-dimensional scan of a brand's fit model to provide a customized avatar for three-dimensional pattern making. The commercialization and availability of three-dimensional scanning equipment in the early 2000s has improved data collection for the formulation of size systems.

Standard body size from public sources

There is not and has never been a universally applied or enforced standard for clothing sizes given the vast differences in body structure throughout the world, even though standardization of measurements may seem to be a solution of fit.

Standard body size from private sources

Some brands such as Levi strauss have embarked upon their own sizing studies with partners, which utilizes radio waves to take body scans through clothing, improving the accuracy of their sizing systems.

9.3.3   Garment fit

Garment fit is influenced by fashion trends, textile characteristics, social and cultural context, function, and demographics. The garment fit depends on ease allowance, and the balance between ease and textile.

Ease

Ease is the amount of difference between the body measurements of the intended wearer and the corresponding measurements of the finished garment, including functional ease and design ease.

Functional ease, or wearing ease, is the minimum required dimensions added to the garment over the body corresponding measurement to allow body movement and comfort. Functional ease varies with the garment category and intended end-use of the garment itself. For example, bicyclists need clothing that is very snug to avoid folds of fabric that catch air and create drag. In general, requirement of functional ease for knit fabrics is significantly less.

Design ease reflects the dimensions added or removed from garment measurements over the corresponding body measurement and functional ease to produce the silhouette desired by the designer. Three-dimensional body scanning is being widely used to observe and measure the air gap between the fabric and the body virtually so as to refine ease allowances.

Fit sessions

Before fitting, the garment will be measured to see if any parts are out of tolerance. During a fit session, a sewn garment is put into a form or a real body and observed by one or more of the product development teams: the technical designer, the designer, and/or the merchandiser. Among other things, the team will review the fit. The evaluation will be a combination of objective and subjective criteria. If the firm has already established accurate standard body size specifications and ease, the fit session will be a test of the design, the patterning accuracy of the vendor, and the relation of the textile to the style. If the firm has not set up standard body measurements and ease, the team will need to correct the fundamental fit in every session.

9.4 Quality assurance

9.4.1 Introduction of production quality control

The production quality process refers to the quality control of the whole manufacturing process from the raw materials entering the factory to the final products. The quality control of production process is directly related to the production of enterprise products, involving planning, supply and marketing, equipment, technology, quality supervision and other departments and the whole production system. The basic principles of production process quality control are "everyone's participation" and "road fortification". The so-called "everyone's participation" means that all personnel related to the production of products are required to actively participate in the quality control activities. The so-called "road fortification" means that monitoring points are required to be set up in each link and process of product production to prevent unqualified products from flowing into the next link or process. Only by carrying out these two basic principles can the quality control of production process be effective.

9.4.2   Contents of quality control in garment production process

As shown in Figure 9.7, the quality control of production process mainly includes the quality control of materials (mainly raw and auxiliary materials), production equipment, production process, design and production process documents (e.g., design and production process change documents), and inspection of garment production. Each production enterprise can combine the characteristics of the product quality requirements of the factory to specify the contents of these links.

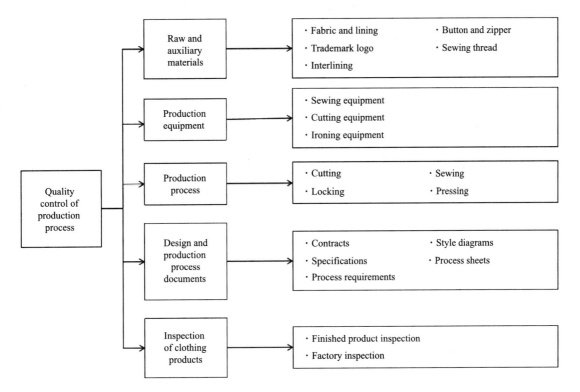

Figure 9.7   Contents of quality control of production process

9.4.3   Several problems in the quality control of garment production process

First of all, the leadership of an enterprise should have a sufficient understanding of the quality control of the production process. They should establish the belief that quality should be grasped from the beginning to the end, and provide strong support for the implementation of production process quality control.

Secondly, we should strengthen the quality awareness of relevant personnel, so that the specific measures of production process quality control can be implemented in the daily work of each worker.

Thirdly, we should improve the post responsibility system related to the quality control of the production process.

Finally, the enterprise management office or the quality assurance office shall be responsible for regular inspection and assessment of the implementation of quality control in the production process, and timely feedback the information to the enterprise leadership.

◎ Summary

Pattern making is a key step to convert design sketches into final product, which will fit the 3D human body. There are several approaches to complete this creative process. Different companies can choose whatever be suitable for them. Before manufacturing, garment fit and specifications will be determined. During manufacturing, product quality will be strictly controlled and several problems related to this process are discussed to reduce drawbacks in the fashion industry.

◎ Key terms

Block patterns (slopers) 原型样板

Design ease 设计松量

Design specifications (design specs) 设计规格

Draping 立体裁剪

Flat pattern method 平面裁剪

Functional ease 功能松量

Fit session 试衣阶段

Grading 放码

Knock-off 抄款

Marker 排料

Mass customization 规模定制

Measurement chart 尺寸表

Pattern drafting 平面制板

Production patterns 生产样板

Quality assurance 质量保证

Sample making 样衣制作

Tag board 标签板

Trace-off (rub-off, reverse engineering) 逆向工程

◎ Product development team members

The following product development team members were introduced in this chapter:

Piece goods/trimmings buyer (面辅料买手): planning the amount of fabric, findings, and trimmings to purchase, determining from which vendors goods will be purchased, and working out financial arrangements for purchases.

Patternmaker (制板师): person who develops patterns for apparel products, including detailed specification lists, or spec sheets, outlining detailed measurement and construction guidelines as well as fabric and trim information.

Quality control manager (质量控制经理): is responsible for the final inspection of garments from the manufacturer in terms of fabric, fit, and construction for quality according to product specification guidelines.

Technical designer (技术设计师): people who interpret the creative design concepts into the development of prototypes that can be manufactured in mass production through accurate product specification completion.

◎ Activities

Imagine that you are working for a company that does knock-offs. Your boss has asked you to start off the design specs for a pair of jeans (Find a pair in your closet) .

1. Make a list of all the pattern pieces that would show up in a pattern chart.
2. Make a list of all the materials. Don' t worry if you don' t know the names of everything. List what you see.
3. Describe the design features, such as the number of pockets, their shapes, specialty in production process, etc.

◎ Review or discussion questions

Compare and contrast the five methods for making patterns. Touch on topics such as brand identity, efficiency, consistency of fit, and creativity.

◎ References

[1] ARMSTRONG H J. Pattern Making for Fashion Design[M]. New York: Harper & Row Publishers, 2000.

[2] BRANNON E L. Designer's Guide to Fashion Apparel[M]. New York: Fairchild Books, 2011.

[3] GLOCK R E, KUNZ G I. Apparel Manufacturing: Sewn Product Analysis[M]. Upper Saddle River, NJ: Prentice-Hall, 2005.

[4] Hu J L. Computer Technology for Textiles and Apparel[M]. Cambridge: Woodhead Publishing Limited, 2011.

◎ Useful websites

[1] https://www.youtube.com/watch?v=bbhlvYjDMg8.

[2] http://www.fitnyc.edu/ccps/certificate-programs/credit/patternmaking.php.

[3] https://www.garnethill.com/size-charts-and-fit-guides/content.

Chapter 10

Production Planning

课题名称： Production Planning

课题内容： 生产策划与产品定价

课题时间： 4课时

教学目的：

1. 理解生产策划的概念和作用。

2. 掌握选择供应商的方法。

3. 了解产品定价的术语和组成部分。

4. 了解产品定价的主要策略。

课前准备： 检索并浏览相关网页，对时装行业的生产策划和产品定价策略有基本认识。

10.1   The role of sourcing

10.1.1   Sourcing strategies

Sourcing is the continuous review of the need for goods and services against the purchasing opportunities that meet quantity, quality, price, sustainability, and delivery parameters, in order to leverage purchasing power for the best value. Companies refer to the combination of resources and vendors they utilize for production of their product line as their **sourcing mix**. A company's sourcing strategy is determined by its strategic plan. Each department brings essential information and expertise to the planning process (Figure 10.1).

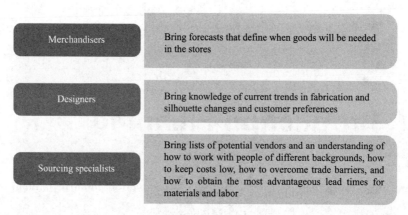

Figure 10.1   Roles in line development

10.1.2   Finding sources

Sourcing specialists can find vendors of materials and labor at more than 200 textile and sourcing shows in over 100 cities in more than 50 countries annually. Messe Frankfurt, an international trade show company, hosts the largest fabric sourcing event in North America, Texworld USA. They also host Intertextile Trade Fairs in Shenzhen and Shanghai, Yarn Expo in shanghai, Techtextile in Frankfurt and Bombay, and so on. Some shows put on by regional trade groups have appealed to international audiences, such as China International Fashion Fair (CHIC). Sourcing professionals may find valuable leads to vendors through their companies' membership in trade organization, such as the American Apparel and Footwear Association, or through networking with member of professional association such as Fashion Group International or China National Garment Association.

10.1.3   Choosing sourcing options

Licensing

For an established brand, a low risky way to add brand extensions to the product mix is to source products through licensing. License provides all services to create and distribute a product, including the design, development, manufacturing, marketing and sales. A brand will license its trade name, design characters and logo under contract and will receive a percentage of the sales.

Sourcing agent

Sourcing through agents is a good option for companies who want to focus their attention on their core competencies of marketing. Sourcing agents provide all the services necessary for production, including sourcing of fabrics and trims, product testing, color matching, sample making, grading, maker making, cutting, garment assembly, finishing, shipping, and export. Agents provide the day-to-day management to supply chain, which may include goods and services from multiple countries and regions. For example, an agent in Hong Kong may facilitate lining from Taiwan, interlining from Hong Kong, and production in Shenzhen. Working with an agent greatly reduces the time needed to establish supply chain relationships abroad.

Case study: Li & Fung

Li & Fung, Hong Kong's largest export trading company, possibly the largest sourcing agent in the world, supplies full package products to Kohl's, Carter's, Target, Benetton Nautica, Roberrto Cavali and many more. With more than 48 trading offices, 63 production countries around the world, Li & Fung is considered as "network coordinators", as they have the responsibility to manage and reconfigure the upstream supply chain to achieve downstream requirements (Figure 10.2).

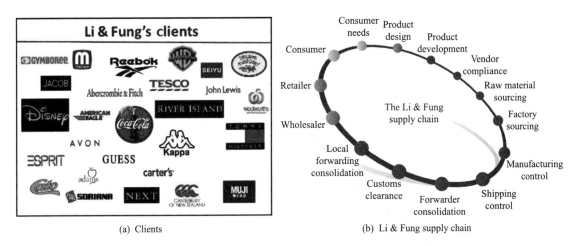

(a) Clients　　　　　　　　　　　　(b) Li & Fung supply chain

Figure 10.2　Li & Fung

OEM (Original equipment manufacturing)

Under OEM contracts, the product development company retains more responsibility for technical planning and design than with full package contracting. It will source materials and create samples, patterns, and markers, as well as provide written standards, assembly instructions, and materials specifications. The vendor sources materials from specified sources while providing cutting, sewing, finishing, and packing. This model has become common with vendors in Bangladesh, Indonesia, Sri Lanka, and Mexico.

CMT (Cut, make, and trim)

Under CMT contracts, the company not only provides the designs, patterns and markers, but also selects, purchases, and arranges delivery of fabrics and accessories. The contractor cuts, assembles and finishes the garments. CMT ensures control of the design, patterning, grading, marking, and materials selection, but risks poor sewing quality and late delivery. As the supply chain in China and Southeast Asia has become more robust, and the level of skills has become increasingly more sophisticated for a low cost, companies tend to outsource the product development and production under full package contracts.

Direct sourcing

The sourcing professional seeks out factory capacity through personal networking, the Internet, or sourcing fairs. The company can select from a range of vendors from CMT factories to full-package facilities. The costs are less because there are no agent fees to pay. USA firms utilize direct sourcing for the majority of their production, including Walmart, Sears, although many of these firms also utilize sourcing agents for a certain portion of their sourcing mix.

Offshore facility and joint ventures

When branded firms build their own offshore production facilities, they are taking advantage of less expensive labor sources abroad while increasing the company's control of the production process, quality, and schedule. However, they take a large amount of risk and responsibility. Initial investment costs are high in obtaining a building, bringing in equipment, and training staff.

10.1.4 Selecting vendors

As shown in Figure 10.3, the following factors are considered when selecting reliable vendors:

- Minimum/maximum capacity;
- Cost;
- Technology;
- Response time;
- Working conditions;
- Compliance infrastructure.

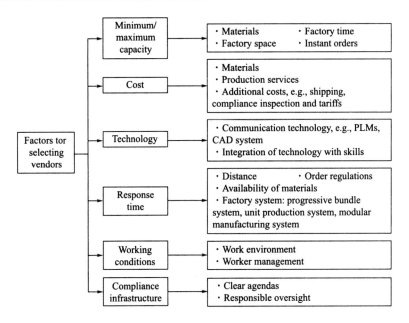

Figure 10.3   Factors for selecting vendors

10.1.5   Business climate

Financial system

Even the most politically stable country can be disrupted economically when outside forces cause current fluctuations or sudden changes in the balance of trade. In developing economies, apparel manufacture and export have a stabilizing influence, as they increase income and attract investment.

Access to market

Manufactures in countries that have well-developed domestic markets tend to have more understanding of consumer needs and have relevant supply chains in place for many material and services. People with relevant skills gained in the local market are available to work in apparel factories for export. Countries with workers who have already been working in export markets have a better understanding of quality standards and cost requirements for foreign clients. Sourcing professionals tend to prefer countries with developed domestic markets and growing exports markets.

Technology penetration

Today the entire global textile and apparel system is dependent on the Internet. PLMs are becoming the standard communication approach. Without real-time connection to customers on the other side of world, the inability to transfer changes in orders rapidly can affect quality and on-time deliveries.

10.2　Costing and pricing

All along the supply chain, each business must pay another business or person for labor and materials, rent and utilities, services, skills, and ideas. These are costs and expenses. The calculations that a business makes in order to earn money to pay costs and expenses plus a profit result in a price.

10.2.1　Retail pricing

Retail pricing depends on marketing and inventory management. The first price on the ticket of an individual item is determined by the strategic plan of the company, its goals and mission, and brand identity. Subsequent prices on the ticket, including markdowns, sales, and promotions, result from decisions made to keep the inventory moving.

Retail pricing formula

A profit and loss (income) statement is a financial reporting tool used to track past performance of income and expenses for a period of time, such as a month, a quarter, or a year. Buyers are given budgets that are based on the financial performance of the store in past seasons and on its goals for the future.

Definition of the terms in the statement:

Gross sales: all of the money earned from sales in a period of time (month, quarter, year).

Return and allowances: money lost when a product is marker down after a return or to resolve a customer complaint.

Net sales: gross sales – returns/allowances = net sales.

Cost of goods sold: the dollar value of unsold inventory from the previous period plus the amount paid to obtain new merchandise.

Gross margin: net sales – cost of goods sold= gross margin.

Operating profit: gross margin – operating expense = operating profit.

Other expenses or income: money earned or spend for activities not related to the retail operation.

Profit before taxes: operating profit ± other expense or income= profit before taxes.

Discount

Suggested retail prices (list prices): is the price at which the manufacturer recommends that the retailer sells the product.

Seasonal discount: wholesale representative may offer a seasonal discount to encourage the retailer to place orders for work to be done at times when the factory is slow. This kind of discount may coincide with times when the retailer needs to attract customers with a special sale or compete with other retailers with similar deals.

10.2.2 Wholesale pricing

The wholesale price is calculated in a similar way to retail. The percentage for cost of goods in the wholesale price will vary greatly depending on the sourcing method used. The cost of goods for a full-package product will be high, but the wholesaler's operating expenses will be low; a CMT cost for the same product will be low, but the operating expenses will be high.

Wholesale price = cost of goods + wholesaler's operating expenses + profit

Although the basic formula for wholesale is similar to retail, there are some significant differences. Wholesale products are sold in bulk with a specific price. The wholesale price may be negotiated with the buyer, but once an agreement is made, the price is fixed for all of the items in the order.

10.2.3 Costing variables for wholesalers

The information needed to calculate the wholesale price will be captured from various sources in one document called the cost sheet. It will be built on information from post-production costing from prior production. New information will be added during preliminary costing and then finalized for production costing. Input will come from members of the wholesaler's team, vendors, brokers, shippers, and government regulations. The process for accumulating information for a cost sheet depends on the kind of vendor. Many wholesalers depend on post-production costing from previous seasons.

Costing professionals will make a decision considering many variables for costing of goods, such as materials, duties and fees, and shipment costs, as shown in Figure 10.4. One decision may

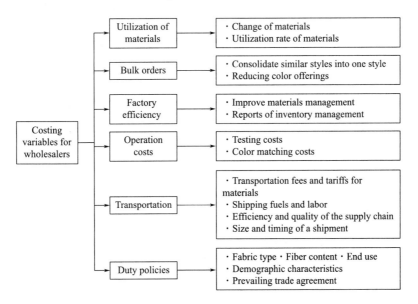

Figure 10.4   Costing variables for wholesalers

impact another. A change of material may mean a higher duty or a longer lead time due to shipping from a more distant supplier. During the process of selecting materials, it is sometimes possible to substitute a fabric or trim that costs less without sacrificing the aesthetic vision of the designer, the brand image, or the functionality of the garment.

10.2.4 Production facility pricing

Direct costing

Production facilities use a variety of methods for calculating operating expense. Their challenge is to earn enough money to cover their costs expense and earn a profit, while not making their prices so high that they are not competitive. Factories that make commodity products, such as T-shirts or jeans, might use **direct costing method** in which all of the operating expense are summed and added as a fixed amount. Then, their own variable costs are material and direct labor.

Absorption costing

Production facilities that produce a wide variety of continually changing products that demand planning, machinery down time for re-tooling, multiple samples, and lots of communication with the client may use **absorption costing**. This method assesses operating expenses as a percentage relative to the direct labor value. This method is based on the assumption that the average assessment percentage is high enough to cover development costs without the necessity of recording the cost of each activity individually.

Activity-based costing

Some companies find that activity-based costing is a more accurate reflection of the cost of each type of product. It measures the cost of each activity (such as administration, marketing, and engineering) and charges the cost of that activity to a product based on its degree of use. For example, a T-shirt would not be charged as much for marker making as a plaid jacket. This method is complicated and time-consuming to apply, and is more apt to be employed by larger businesses with multiple factories.

10.3　Production

10.3.1　Cutting instructions

The tech pack should include cutting instructions to alert the contractors who make markers or spread and cut fabric within unique requirements of the fabric or style. The layout of pattern pieces in relation to a printed or woven design has to be specified. Some companies provide a list of pattern pieces to be sure that all internal components such as interlinings and chest pieces are included.

Listing every pattern piece can limit the flexibility of the factory. A cut order plan will be generated by the production department. It will confirm the exact style, sizes, and quantities that are to be cut to complete the contracted order.

10.3.2   Assembly instructions

Three major production assembly methods or systems are employed in apparel manufacturing plants.

Progressive bundle system

The progressive bundle system utilizes traditional assembly line concepts. The bundles of cut parts are transported to the sewing section and given to the operators scheduled to finish the operation. Thus bundles may be handled from one sewing station to another in various forms such as tied bundles, bags, pocketed bags, bundle trucks, boxes and baskets, etc.

Unit production system

In the unit production system, garments are introduced to each operator's station via an overhead transporter. Transporter moves the developing garments from station to station, and the operator attaches the newest component or performs the appropriate tasks on each garment as it reaches the station (Figure 10.5).

(a) Progressive bundle system   (b) Unit production system

Figure 10.5   Production system

Modular manufacturing method

Modular manufacturing methods group the operators into teams or modules, and each team focuses on efficient completion of each garment moving through its module by passing it off to the next station in the module as the tasks for that station are completed, as shown in Figure 10.5(b). This newer production method focuses on team efforts and utilizes in-process quality assessment whereby corrections are made as errors occur rather than at the end of the production line, when some errors cannot be repaired.

10.3.3　Finishing

The last steps of apparel production are referred to as **finishing**. The process includes trimming threads, final inspection, repairing defects, pressing, folding, and packing. Most garments require finish pressing. This is done in addition to any in-process pressing that might have been required to achieve desired effects within the construction process.

The fold and pack operation includes the addition of accessory items such as belts and hangtags. Garments are then folded or hung on hangers as outlined in the original specifications. The individual garments may be placed in plastic bags for protection, or they may be grouped before being placed in protective covering to prevent soiling during transportation. Finally, the garments are placed into cartons or placed on hanging racks shipment.

◎ Summary

Whether a sourcing specialist is looking for replacement of current vendors or starting a new venture, he or she should do some research. Selection of a vendor cannot stop with evaluation of the factory's characteristics alone. Traditional costing has been based on costs that related directly to the business conducting the activity. Buyers and managers use accounting tools to determine prices that will keep the inventory moving out of the store. The steps of apparel production include scheduling, spreading and cutting, assembly, finishing, and distribution to retail venues.

◎ Key terms

Absorption costing 归纳成本计算法
CMT 来样加工
Costs and expenses 成本和费用
Cost of goods sold 销货成本
Direct costing 直接成本法
Direct sourcing 直购
Fixed costs 固定成本
Gross sales 销售总额

Net sales 净销售额
OEM 贴牌加工
Preliminary costing 初步成本
Production cost 生产成本
Return and allowances 退货及折让
Sourcing agent 采购代理
Sourcing mix 采购策略
Wholesale price 批发价

◎ Product development team members

The following product development team members were introduced in this chapter:

Production planner(生产策划师): is responsible for planning, anticipating all of the parts needed to make the final product.

Sourcing manager (采购经理): is the individual responsible for sourcing production.

◎ Activities

Choose one fashion brand, try to visit the store and surf online and analyze the sourcing and pricing strategies, comparing with one of the competitors.

◎ Review or discussion questions

1. If you were starting your own line of clothing, how would you find sources for your production?

2. Think of a favorite retail store that you visit frequently, try to remember the various pricing strategies they use. Do these pricing strategies work for you?

◎ Semester project

Semester project IV: Select vendors

1. Imaging you as a source specialist, define the type of sourcing options.

2. Try to find a vendor for your option using one of the methods in this chapter, based on research results.

◎ References

KEISER S, VANDERMAR D, GARNER M B. Beyond Design, The synergy of Apparel Product Development[M]. 4th Edition. New York: Fairchild Books, 2017.

◎ Useful websites

[1] https://www.bilibili.com/video/BV1KM4y1N78f.

[2] http://www.sjfzxm.com/global/en/page.html.

[3] http://www.doc88.com/p-2055614753398.html.

[4] http://www.garmentmanufacture.com/.

Part IV
Distribution Planning

Chapter 11

Distribution

课题名称： Distribution

课题内容： 分销

课题时间： 4课时

教学目的：

1. 理解分销策略的重要性。

2. 描述分销渠道的不同形式。

3. 讨论选择不同分销渠道的基准。

4. 辨识直接和间接销售。

5. 理解批发和零售在分销中的角色。

课前准备： 浏览本章提供的相关网页，增进对各公司分销渠道的感性认识。

11.1 Distribution plan development

11.1.1 Distribution strategy

Distribution encompasses all the physical activities necessary to make a product or service available to the target consumers when and where they need. The distribution strategies and channels that a company chooses should match the characteristics of their product. Prior to positioning product in the marketplace, product developers define an effective multichannel strategy for selling. They must know their product and intended market segment. To appropriately target distribution channels the best barometer is the price classification. Targeting retailers that sell products in the same price category is a strong indicator for distribution partnership. By reviewing the market and determining the most appropriate retailers—from brick-and-mortar to online, a company can create a multichannel strategy to reach the maximum number of customers in its price classification segment.

To visit stores and check the distribution of the competition will often aid in knowing which retail stores are viable for partnership. Nowadays, distribution can no longer be viewed as a random, standalone, or fixed activity. Companies are constantly researching, identifying, and evaluating multiple channels of distribution to determine which yields the maximum sales for the product. The inherent agility of a strategy map is necessary.

A distribution strategy is a plan that provides specific direction to how a company will sell its product to reach the end user. It sets direction for positioning goods into the marketplace. A distribution channel (or a marketing channel) is a path through which goods flow from a company to its customer. In some cases, a channel may simply be the producer carrying out the distribution functions. Other distribution channels consist of the producer plus one or more resellers. In these channels, the resellers handle some of the distribution functions. The resellers in a distribution channel are called intermediaries. An intermediary, or middleman, is an independent business that specializes in linking

Transactional functions
Buying: purchasing products in order to resell them Selling: promoting products to potential customers and soliciting orders Risk taking: assuming business risks by owning goods that can be damaged, or become obsolete

Logistical functions
Concentration: bringing goods from various places together in one place Storing: maintaining inventories and protecting goods in a way that meets customer needs Sorting: purchasing in quantity and breaking into amounts desired by customers Transporting: physically moving goods from where they were manufactured to where they are to be bought

Facilitating functions
Financing: providing credit or funds to facilitating a transaction Grading: inspecting products and classifying them into categories based on quality Market research: gathering data and reporting information on market conditions, expected sales, consumer trends, and competitive forces

Figure 11.1 Functions performed by intermediaries

sellers with consumers. Figure 11.1 summarizes the functions performed by intermediaries.

Figure 11.2 shows examples of distribution channels.

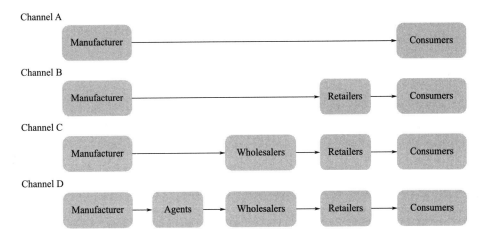

Figure 11.2   Examples of distribution channels

In channel A, there are no intermediaries, and goods move direct from producer to consumers. This type of channel organization is called direct marketing.

In channel B, goods move from producer to retailers to consumers. A retailer is an intermediary that businesses to consumers (B2C).

Channel C is the most common channel for consumer goods. In this channel, the producer sells to wholesalers, which sell to retailers, which in turn sell to consumers. A wholesaler is an intermediary that sells to other intermediaries (B2B).

Channel D is the most indirect channel. Goods pass from producer to agents to wholesalers to retailers and then to consumers. Unlike other types of intermediary, an agent negotiates purchases or sales but does not take title to the products it handles. Agents coordinate a large supply of goods when there are many small manufacturers and retailers. Small manufacturers that lack the capital for their own sales force use agents, often called manufacturer's representatives, to serve as an independent sales force to contact wholesalers.

11.1.2 Managing distribution channel

The process of channel management begins with selection of a type of channel and of the specific organizations that will distribute the product. The right channel members will know how to get the product to places where the target market will buy it. In contrast, with the wrong channel members, interested buyers might never encounter the product. Although it is crucial for a producer to select a distribution channel, other channel members also make decisions about which sellers to represent and which target markets to serve. As summarized in Table 11.1, marketers should base the selection of a viable distribution channel on characteristics of the target market, the organization's marketing objectives, the nature of the product itself, the intermediaries, and the marketing environment.

Table 11.1 Considerations for selecting a channel of distribution

Channels	Considerations
Target markets	Number Geographic dispersion Purchasing patterns Susceptibilities to different selling methods
Marketing objectives	Efficiency Intensity of distribution
Product characteristics	Bulkiness Stage in life cycle
Intermediary characteristics	Availability Willingness to accept product or product line Strengths and weaknesses
Marketing environment	Competitors Economic conditions Laws and regulations

Once a distribution channel is in place, the members of that channel must cooperate to get the product to market. Channel members must strive to achieve the general goal of distributing the product profitably. However, the conflict between members of the distribution channel is common. Conflicts involve such issues as how to allocate profits, which channel members will perform which services, and which channel members will make certain decisions about marketing the product. Beyond the tedious task of creating a distribution plan and determining the most viable channels, a company must also develop channel management. Channel management develops the policies and procedures that will be adopted in order for all members involved in the distribution channel to perform the required functions. Good communication, trust, and understanding among all members, along with contracts are key to the effective functioning of the channel.

11.2 Direct market distribution

A direct market channel is any conduit that connects the producer of a product to the end user. The connection is a business-to-consumer relationship (B2C), and there is no intermediary involved in all situations. Direct market distribution allows the company the opportunity to develop relationships with the product's target market, thus can provide insightful interaction and feedback to respond to the needs and wants of the customers. Commonly used direct market distribution strategies include direct sales force, e-commerce websites, direct mail, and other one-to-one techniques to communicate and sell to the customers and clients. Nowadays, many companies are using direct market distribution exclusively or in concert with other marketing channels.

11.2.1   Direct sales force

Direct selling

Direct selling, a face-to-face distribution approach, is one of the oldest distribution methods. It is the most traditional direct market channel. Away from a fixed retail location, an independent sales representative sells a product to a consumer in person. The direct sellers are not employees of the company who develop the products. Instead, they are independent contractors who market and sell the product of a company in return for a commission on the sales. Orders are usually placed in person, via telephone, or via the salesperson's website. For fashion products, make to measure (MTM) employs this approach to reach their target customer (Figure 11.3).

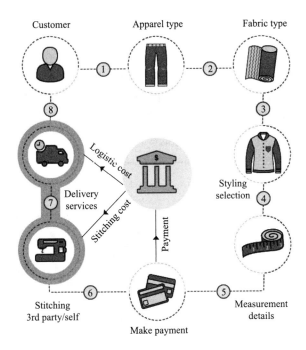

Figure 11.3   Online custom tailoring platform

Corporate sales associates

Factory stores or industry tours become more and more popular in China. Many apparel factories or companies run brick-and-mortar retail operation, build retail environment, to sell products developed by themselves, and even from other.

11.2.2   Internet market distribution

The internet has revolutionized direct marketing for promoting the sale of products and services to the targeted audience. E-commerce, also known as electronic commerce or internet commerce, refers to the buying and selling of goods or services using the internet, and the transfer of money and data to execute these transactions. Since the first ever online sale: on August 11, 1994 a man sold a CD by the band Sting to his friend through his website NetMarket (an American retail platform), e-commerce has evolved to make products easier to discover and purchase through online retailers and marketplaces.

Company website

With the novelty of customizing websites and e-commerce storefronts channels, companies can eliminate geographic limitations because people around the world have the same access as the person across the street. Utilizing a company website as a direct market distribution tool can be more targeted, more flexible, more responsive, more affordable, and potentially more profitable in the long term. E-commerce has become a viable fit for direct marketing.

Social media

Social media is another viable and powerful direct market channel. Large companies use it to communicate exclusive offers, generate sales, and create a community of loyal brand advocates. Figure 11.4 shows two examples to promote products through WeChat and opt-in E-mail. These two channels allow companies to get a more complete view of customers, relevant goods to deliver, effective communications.

Figure 11.4　Promotion and sales through social media

11.3   Indirect distribution (B2B)

Indirect distribution is when a manufacturer utilizes intermediaries to sell their products, products are sold from business to business, commonly called B2B. As illustrated in Figure 11.2, an indirect channel can be as simple as using a retailer to get to the end consumer (channel B), or it might imply using both a wholesaler and a retailer (channel C), or even an agent, a wholesaler, and a retailer. The first channel is also classified as a single-party system because only one intermediary involved. The latter two channels are classified as a multiple-party system because they have the product passing through two or more distributors before reaching the target customer.

11.3.1   Wholesalers

A wholesaler is a business that is primarily engaged in buying, taking title to, usually storing and physically handling goods in large quantities, and reselling the goods (usually in smaller quantities) to retailers or to other wholesalers. Although most consumers have little contact with wholesaling activities, wholesaling is a major business.

Wholesalers perform some of or all of the distribution functions listed in Figure 11.2. For example, they may transport and warehouse goods, exhibit them at trade shows, and tell store managers which products are moving the fastest. Stores that carry a wide selection of merchandise often like the convenience of dealing with wholesalers, which make many different products available from a variety of sources. On the other hand, producers may appreciate such benefits as expertise in locating retailers or other organized buyers. Figure 11.5 shows a website bridging between wholesalers and retailers.

Figure 11.5   Bridge between wholesalers and retailers

11.3.2 Retailers

Retailers sell products directly to target consumers. Retailers may buy from either producers or wholesalers. The most obvious examples of retailers are stores, but any organization in the business of selling to consumers is engaged in a form of retailing, whether or not it operates a store. Retailing is even bigger business than wholesaling.

Retailers provide a convenient way for producers to make their products available to consumers. From the consumer's perspective, retailers offer a variety of benefits. They make goods available at convenient hours and during the desired seasons. Retailers make products easier to purchase by accepting credit cards, apple pay, Ali-pay, Wechat pay and so on. They also may modify products to make them more fit to consumers, like modifying the length of pants.

11.3.3 Attending trade shows or sales meetings

In today's apparel environment, a majority of textile products and apparel styles are not produced in quantity until it is known which of these products are wanted by buyers for presentation to consumers in the marketplace and in what volume they are to be produced. In the case of private label merchandise, retail merchandising staff, working with product development staff, determine which products from the styles presented by the designers have the greatest sales potential. They then narrow the offerings to a selection, called their line or unit merchandise plan, that fits the firm's overall plan for the coming season. They determine the size range, colors, and quantity of each style to be offered for sale in their stores.

Trade shows

Fashion product developers can participate in seasonal trade shows to introduce new product lines to retailers. Trade shows are places where manufacturers of products sell to retailers or wholesalers. Trade shows are the foundation of the textile and apparel sales calendar. Referred to in the business as simply a market or market weeks, they are held in major cities throughout the world at carefully scheduled times throughout the year. They are often set up for specific categories of merchandise such as men's, bridal, children's, or the most visible category, women's fashions. In the United States, for decades New York City has been the U.S. venue of choice for major fashion markets, especially in women's wear. Figure 11.6 shows two famous trade shows: Premier Vision in Paris and CHIC in Shanghai.

Sales meetings

For branded goods and in more traditional manufacturing settings, an additional step must be taken before garment samples are put into full production. That step is selling the designs to customers, in

Premier Vision in Paris   CHIC in Shanghai

Figure 11.6   Trade shows

this case, usually retail buyers. At this point, only samples have been produced to aid in selection of contractors and for sales representatives to use for their sales responsibilities. Sales representatives take original production samples to sell to buyers at major market centers and in wholesale showrooms. As orders are taken and processed, styles that sell well enough to be deemed profitable are put into full production. Those styles that are not selling well are typically pulled from the line before any more production capacity is devoted to them.

Line presentation

To attend trade shows, or other sales meeting, product developers need to make some documents to present their new product lines. A **line sheet** is a set of documents that communicate key product information to prospective buyers. Since they are used to assist buyers in writing product orders, they should be detailed so that a buyer does not need to see the actual product. Normally, line sheets include:

- Product season, name and description;
- Style numbers together with product images;
- Size range;
- Wholesale price and suggested retail price;
- Color and fabric information;
- Delivery dates and order cut-off dates;
- Order minimums by item and/or total dollar amount;
- Company and/or sales rep contact information.

Figure 11.7 shows an example of a line sheet. Nowadays, many companies will provide a hard copy or provide an electronic file on a USB drive.

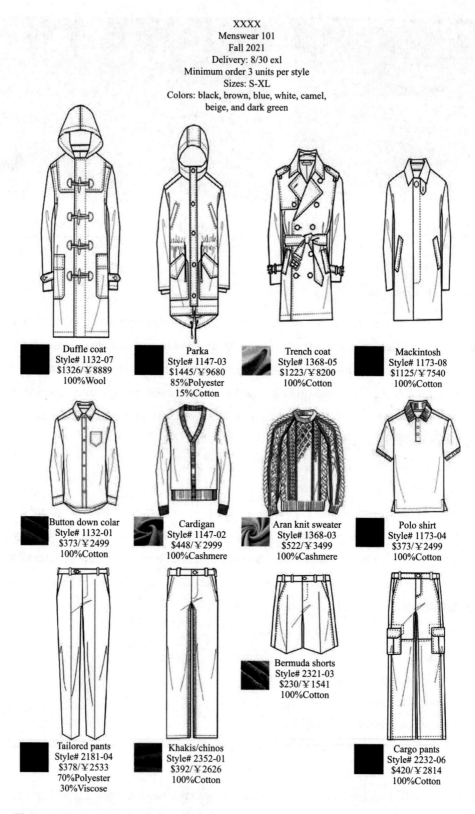

XXXX
Menswear 101
Fall 2021
Delivery: 8/30 exl
Minimum order 3 units per style
Sizes: S-XL
Colors: black, brown, blue, white, camel,
beige, and dark green

Duffle coat
Style# 1132-07
$1326/¥8889
100%Wool

Parka
Style# 1147-03
$1445/¥9680
85%Polyester
15%Cotton

Trench coat
Style# 1368-05
$1223/¥8200
100%Cotton

Mackintosh
Style# 1173-08
$1125/¥7540
100%Cotton

Button down colar
Style# 1132-01
$373/¥2499
100%Cotton

Cardigan
Style# 1147-02
$448/¥2999
100%Cashmere

Aran knit sweater
Style# 1368-03
$522/¥3499
100%Cashmere

Polo shirt
Style# 1173-04
$373/¥2499
100%Cotton

Bermuda shorts
Style# 2321-03
$230/¥1541
100%Cotton

Tailored pants
Style# 2181-04
$378/¥2533
70%Polyester
30%Viscose

Khakis/chinos
Style# 2352-01
$392/¥2626
100%Cotton

Cargo pants
Style# 2232-06
$420/¥2814
100%Cotton

Figure 11.7   An example of a line sheet

◎ Summary

Product placement is equally as important as the product development. The distribution strategies and channels that a company chooses should match the characteristics of its products. Two significant methods of placing product into the market for consumption are executed by direct and indirect market distribution strategies. Multichannel distribution strategies involve offering merchandise through more than one type of distribution channel. To attend trade shows, or other sales meeting for wholesale, product developers make line sheets to present their new product lines.

◎ Key terms

Barometer 晴雨表

Brick-and-mortar 实体店

Catalogs 产品目录

Delivery 发货

Direct market distribution 直接市场分销

Direct sales 直销

Distribution channel 流通渠道

E-commerce or e-tailing 电子商务

Facilitating 辅助、促进

Indirect market distribution 间接市场分销

Indirect sales 间接销售

Intermediary 中间商

Line presentation 系列展示

Line sheet 系列清单

Logistical 物流的

Multidimensional customers 多元消费者

Non-store retailing 无店铺零售

Order 订单

Sales meeting 展销会

Sales representative 销售代理

Social media 社交媒体

Trade shows 贸易展览

Transactional function 事务功能

Transporting 运输

◎ Product development team members

The following product development team members were introduced in this chapter:

Buyer（买手）: an individual who is responsible for all of the product purchases for a particular department of a company. His or her responsibility includes: monitoring fashion trends and determining which items their customers will buy at a profit, locating suppliers and negotiating prices, shipping, and discounts, using fashion sense, knowledge of trends, and understanding of target customers to create desirable merchandise assortments for the store.

Planner（策划者）: works in collaboration with a buyer to develop sales forecasts, inventory plans, and spending budgets for merchandising.

Sales representative（销售代理）: an individual who is responsible for selling the product line to retail customers. The sales representative may be independent or manufacture representatives. They typically have a territory and sell products in market centers and at trade shows.

Vice president of marketing（市场营销副总裁）: an individual that oversees all distribution channels. He or she manages marketing, customer service, sales, retail merchandise associates. He or she pays close attention to current retail business trends and how they will affect the company.

◎ Activities

1. Find examples for each type of distribution channels.

2. Identify some examples of direct selling in fashion business.

◎ References

[1] CARR M G, NEWELL L H. Guide to Fashion Entrepreneurship: The Plan, the Product, the Process[M]. New York: Fairchild Books, 2014.

[2] CHURCHILL G A, PETER J J P. Marketing, Creating Value for Customers[M]. Austen Press, 1995.

◎ Review or discussion questions

What distribution strategy will help your business grow? Explain why.

◎ Semester project

Semester project V: Develop a distribution plan for your brand

1. Select at least two direct and two indirect market channels that you would consider viable for your new products, and discuss how these channels will align and facilitate to reach your target consumers.

2. Find three retailers for your product, explain why your brand would be best suited for these three retailers.

◎ Useful websites

[1] Alibaba: https://www.alibaba.com/premium/clothing_suppliers.html.

[2] Apparel News-Trade Show Calendar.http://apparelnews.net/events.

[3] Brands Distribution. https://www.brandsdistribution.com/en/.

[4] Buyer at Large.www.buyeratlarge.com.

[5] Buyerly.http://buyerly.com.

[6] GRIFFATI, Wholesale Fashion Brands. https://www.griffati.com/.

[7] Electric Retailing Association: www.retailing.org.

[8] How to create a distribution strategy that actually makes money. http://www.thomholland.com/distribution-strategy.

[9] RetailSales Connect. www.retailsalesconnect.com.

[10] shop.org of the National Retail Federation.www.shop.org.

[11] U.s.Small Business Administration.https://www.sba.gov/business-guide/grow-your-business/expand-new-locations.

[12] https://wenku.baidu.com/view/bbb051f3f90f76c661371a60.html#.

Chapter 12

Marketing Communications

课题名称：Marketing Communications

课题内容：营销传播

课题时间：2课时

教学目的：

1. 了解市场行销者如何和目标市场沟通。

2. 掌握各行销传播渠道及其功能。

3. 能辨别营销传播组合的各元素。

12.1   Communications mix

The marketing mix is a marketing tool to satisfy customers and company objectives. It facilitates putting the right product in the right place, at the right price, at the right time, and with the right promotion. Product, price, place or placement, and promotion (4Ps) constitute the marketing mix. Product, price and placement have been introduced in previous chapters. Promotion is now being renamed marketing communications to which perhaps reflect its growing importance and increased profile within the marketing mix.

Communication is the transmission of a message from a sender or a source to a receiver. It is a process. The source encodes the message, converting it into a group of symbols that represent images or concepts. Through a medium of transmission, such as television or the internet, written words, photographic images, and musical sounds, the source sends the message to the receiver, the individual or group for whom the message is intended. The receiver then decodes the message, converting symbols back to the original images or concepts contained in the message. In successful communication, the sender and receiver share the same meaning of the message.

Before consumers or organizational buyers can purchase a product, they need to know about it. They need to know what the product is, how it will benefit them, and where they can find it. Providing this information is the goal of marketing communications. Marketing communications involve communicating with target markets to inform them about a product and influence them to buy it. As summarized by the AIDA model, the message must achieve first Attention, then Interest, Desire, and Action. The communications mix blends together several different elements to create the overall strategy for marketing communications. The elements of this mix for fashion products may include advertising, public relations, sales promotion, personal selling.

12.2   Advertisement

When people think of marketing messages, usually advertisements first come to their minds. Advertising is paid, nonpersonal communication through various media by organizations that are identified in the message and seek to inform and/or persuade members of a particular audience. The media often used by fashion companies are screen, print, outdoor (billboards). Each medium has its advantages and disadvantages, and the medium that is most appropriate for one product or brand may not be suitable for another.

Screen–based advertising

Television advertising is the most expensive method of advertising because it reaches the maximum number of people. However, for fashion product developers and retailers it is not always cost-effective as the fashion consumer or the target market may not see the advertisement. Also, it may be lost among other non-fashion advertisements, and a short three-minute commercial cannot always show the full range of items available. However, television advertising can be a useful medium for brand image creation. For many fashion companies, targeting specific programmes which are likely to be watched by the target market may give some precision and an opportunity for the promotion of international brands such as Levi's.

Another option to promote product on-screen is to pay to promote a product within the content of a television show or movie, this is called product placement. Increasingly movies and television shows are boldly zooming in on brand logos to promote the products used on film. This form of promotion works best for brands with high name recognition and status. The well-known movie *the devils wear Prada* is a good example. By this approach, Korean drama has been proved to be successful in promoting the Korean-style fashion.

In an attempt to re-engage the customer, some retailers are adding television screens to their selling space. Approaches range from the use of small 4-inch monitors placed on the shelf next to the merchandise to large plasma TVs designed and placed to raise the status quotient. They may play anything from a 30-second commercial to a 15-minute loop of a runway show (Figure 12. 1).

The devils wear Prada   Korean drama: *The heirs*

Figure 12.1   Placement advertisement

Print advertising

Fashion product developers and retailers have traditionally relied on print mediums as opposed to TV/broadcast mediums to get the attention of customers. Print advertising allows consumers to select the advertisements that most interest them and thus are considered less invasive. Many product developers and retailers see advertising in magazines as the most effective means of communicating with a specific demographic, as it can be fine-tuned to the target market of the magazine. Magazines provide media packs, which outline their target market and pricing strategy. The media pack gives demographic details on the age profile, social classification, income level and education of its core readership, and a lifestyle profile is provided for their potential advertisers. For example, *Vogue* is renowned as a style bible all over the world and has a very clearly defined target market: "To be in *Vogue* is to be in fashion". The *Vogue* reader "is style conscious … enjoys an affluent, active lifestyle", dining out, pursuing cultural activities, taking frequent holidays and, of course, shopping at upmarket stores, and spending on clothes（Figure 12.2）.

Vogue   *Elle*

Figure 12.2   Fashion magazines

Outdoor advertising

Ambient media denotes advertising on billboards, the Underground, street furniture, taxis, etc. It is relatively inexpensive, and its main advantage is that it can be highly targeted to transport users within the area of the fashion stores. A disadvantage is that it cannot give a lot of information as people will only have a moment to glance at it, and it must not be too distracting in content as it could cause people to lose concentration（Figure 12.3）.

Figure 12.3   Outdoor advertising

12.3   Public relations

The Institute of Public Relations (IPR) simply defines PR as: PR is about reputation—the result of what you do, what you say and what others say about you. Public relations may include an organization or individual gaining exposure to their audiences using topics of public interest and news items that do not require direct payment. This differentiates it from advertising as a form of marketing communications. The aim of public relations is to inform the public, prospective customers, investors, partners, employees, and other stakeholders and ultimately persuade them to maintain a positive or favorable view about the organization, its leadership, products, or political decisions. In essence, PR can be seen as managing the corporate identity.

In fashion marketing, PR is a very effective method of promotion. PR can simply be getting goods into the public arena at one end of the spectrum, or at the other extreme, refuting any adverse publicity that the company may face. Therefore, it is important for a company to have a PR department in place in order to be prepared for adverse reactions from the press or public. PR, if used effectively, can be seen as the guardian of the corporate reputation. In fashion marketing, PR may be done by an in-house department, or by an external agency. The aims of PR for fashion businesses can be:

- Raise or confirm the profile of the brand/retailer;
- Place products in the public arena;
- Enhance other parts of the promotional mix;
- Communicate with influential media.

Celebrity endorsement

PR also has an important role to play in the use of celebrities by brands. Many consumers are fascinated by which celebrities attend runway shows and what the celebrities wear on the red carpet at award shows. Their choices are quickly reproduced for the mass market. Celebrity endorsement is a particularly powerful form of promotion and is increasingly being used by companies. Linking a well-known personality with the brand can give many benefits to a brand that advertising alone cannot give.

If a well-known personality endorses a product or wears a brand through a sponsorship deal, the public may well understand that they are being paid to promote the product. However, the effect is one of credibility, particularly if the personality is admired, in which case they imbue the product with their attributes of good taste, attractiveness, etc. Sportswear brands have used celebrity endorsements to great effect in matching the brand with the personality. However, the major disadvantage of celebrity endorsement is if the endorser "falls from grace", by being accused of some scandal or wrong doing, then the product with which they are linked will inevitably suffer a similar fate, and can become tainted with the same public disapproval.

Sponsorship

Supporting causes, sponsoring special events, and inviting prized customers to special events have all become important methods of marketing. The sponsorship of special events has been around for a long time. Sponsorship of the right event can increase a brand's cool factor while also reserving rights for merchandise distribution and promotion.

12.4　Sales promotion

Sales promotion is used to promote demand growth and increase the sales of certain products. The purpose of promotion is to stimulate additional consumption. This method of promotion revolves around offering a discount on merchandise. There are various ways for sales promotion, for example, coupons, discounted sales, free sample products, additional benefits or services per purchase, or offering gifts. The cost of this obviously requires budgeting into the original promotional strategy, but provides a method of immediate feedback on the success of the promotion and would perhaps turn new or occasional shoppers into regular purchasers (Figure 12.4).

Figure 12.4　Various sales promotion

Owing to the seasonal nature of fashion, mid-season and end-of-season sales are the most frequent methods of sales promotion and an effective way of reducing the stockholding in order to make space for new merchandise; however due to the rise of the "fast fashion" phenomenon, which is the consumers' constant demand for newness, stores are often in continual markdown.

12.5   Personal selling

Personal selling is selling that involves personal interaction with the customer. Salespeople can be persuasive and influential, and two-way communication allows for questions and other feedback. In fashion marketing, the role of the sales assistant generally varies according to the type of outlet. Fashion outlets for the youth market tend to be self-service as the "fashion enthusiasts" are quite happy to select, try on and purchase garments without much help or interference. However, the older consumer may require a personal service in terms of advice and alterations. Therefore, the type of sales assistant recruited by a fashion company tends to attract and reflect the target market in terms of age, size, demographics and lifestyle.

Upmarket retailers offer personal shopping service, which is now increasingly a feature of the mass market. A selection of merchandise is provided from which the customer can make a selection in some privacy and comfort. It is also popular for the busy, time constrained, working executive or for a special-occasion purchaser. As in the cult of celebrity, everyone wants their own personal stylist and stores are catering to this demand by providing a stylist service.

12.6   Managing the communications mix

As at every other level of planning, the marker needs to set objectives for communications. The objectives must be clear, specific, and challenging but achievable. Communication objectives should also support the overall marketing objectives. If the marketing objectives include a particular level of sales, the communications objective should include reaching enough potential buyers to generate that level of sales. If the marketing objectives involve positioning the product as a prestige item, the message should convey that image.

To achieve communications objectives, instead of functioning alone, the elements in the communication mix need to work together. To select the right mix of communications elements, the marketer evaluates the contributions that each element can make to achieving communications objectives, as well as the overall marketing objectives. The marketers can ask a sample of the target market to view a prototype of the message and report their reactions. They might do so in conjunction with evaluating the product itself. The product's stage in its life cycle also needs to be considered.

The marketing manager must prepare a budget for marketing communications. While the marketer is establishing objectives for the communications plans, thinking about budget has already begun. When the marketing manager takes into account all issues related to selecting a communications mix, as well as the cost of the various elements, the budget begins to come into focus. There are some common budgeting techniques marketers can use. Among them, a simple to use one is the percentage of sales method. It is based on a specified percentage of either actual or estimated sales.

To evaluate the communications effort, feedback is an important means. Even before a communication plan is fully implemented, marketers seek feedback in the form of evaluations from a variety of sources, including marketing researches, experts familiar with the product, a sample group of the target market. Once the effort is ongoing, the marketing team uses a number of tools to monitor it and make adjustments as needed.

◎ Summary

Through marketing communications, companies seek to inform their target markets about products and influence their attitudes and buying behavior. Effective marketing communications may provide customers with information and make them aware of the product, increase demand for the product, distinguish the company's product from its competition, favorably bias the customer toward the product, and stabilize product sales. To achieve communications objectives, the communication mix blends several elements: advertising, public relations, sales promotion, personal selling. To select the right mix of elements, the marketers consider the nature of the target market, the nature of the product, the product life cycle and budgeting.

◎ Key terms

AIDA model 爱达模式

Ambient media 环境媒体

Billboard 广告牌

Celebrity endorsement 明星代言

Communications mix 营销传播组合

Discount 折扣

Cupon 优惠券

Personal selling 人员推销

Print advertising 印刷广告

Product placement 植入式广告

Public relations 公共关系

Sales promotion 促销

Screen-based advertising 屏幕广告

Sponsorship 赞助式广告

Street furniture 街头公共设施

Zero in 瞄准

◎ Product development team members

The following product development team members were introduced in this chapter:

Marketer（营销员）: the marker sets objectives for communications and selects what elements to include in the communication mix. The marketer must also prepare a budget for marketing communications. When the effort has begun, the marketer must evaluate its success and make adjustments as needed.

◎ Activities

1. Choose a favorite brand, identify what communications mix this brand often uses.

2. Visit local stores and identify the modern technological installations different brands utilize in their store to engage customers. Think about what it is and why they are using it. Discuss your observations and insights in class.

◎ Review or discussion questions

1. Why sales promotion is successful only as a short-term effort?

2. Identify some new directions in fashion marketing communications.

3. In what ways could you generate good publicity about a sports brand store and bring in new customers?

◎ Semester project

Semester project VI : Develop a marketing plan for your brand

Set a strategy to market your product to the ultimate customer. Create at least one objective, strategy, and tactic per phase of business for your marketing plan.

Once the marketing mix has been developed, answer the questions below:

1. Does the product meet the customer's needs or wants? How?

2. Where will they purchase the product? What percentage will be online versus at a brick-and-mortar location?

3. Is the price competitive? What brand is the price leader?

4. What social media channels are being used for product introductions?

◎ References

CHURCHILL G A, PETER J P. Marketing, Creating Value for Customers[M]. Austen press, 1995.

◎ Useful websites

[1] Advertising Standards Authority Ltd. www.asa.org.uk.

[2] Campaign. https://www.campaignlive.co.uk/.

[3] BrandChannel. www.brandchannel.com.

[4] "Free sample Business Plans" Bplans.www.bplans.com/sample_business_plans.cfm.

[5] "How to write a marketing plan" Entreprener.com.www.entrepreneur.com/article/43018.

[6] "Write Your Business Plan". U.S.Small Business Administration.https://www.sba.gov/business-guide/plan-your-business/write-your-business-plan.

Chapter 13

Retailing

课题名称： Retailing

课题内容： 服装零售

课题时间： 4课时

教学目的：

    1．理解服装零售的概念及市场策略。

    2．能定义视觉营销。

    3．能理解促销的手段。

    4．了解零售管理方法。

课前准备： 检索并浏览相关服装零售的技巧、策略等网站、书籍，形成对服装零售的基本认知。

13.1 Retail marketing mix

13.1.1 Definitions

Retailing is a distribution channel function where one organization buys products from supplying firms or manufactures the product themselves, and then sells these directly to consumers. In the majority of retail situations, the organization from which a consumer makes purchases is a reseller of products obtained from others and is not the product manufacturer. The business of buying clothes from manufacturers and selling them to customers is known as **fashion retailing**.

The **retail marketing mix** can be defined as the composite of all effort which is programmed by management and which embodies the adjustment of the retail store to its market environment. Retail marketing strategy is closely related to the general marketing strategy, however the standard "4Ps" marketing framework (product, price, place and promotion) is not sufficient as it concerns itself with product; for the retail industry the focus needs to be on market orientation which requires a more detailed framework, and therefore an adaptation of the 4Ps model is developed. There are a number of specific elements to the retail mix that have been detailed by academics, and these will be discussed in this section.

13.1.2 Place

Someone thinks the most critical element of the retail mix is place or location. Store in the right place with other similar retailers impacts traffic and customer types and therefore similar retailers will benefit from being located near to each other. Making location decisions is critical to the success and growth of retailers. However, McGoldrick states that a good location will not suffice and certainly does not make up for other areas of the retail mix that are weak. Interestingly in a time when many consumers are focused on price, it could well be this factor that influences where consumers shop. Some of the value retailers might not always locate in prime locations, but is still successful in attracting visitors despite their store design, atmospherics and overall shopping experience.

13.1.3 Product

Fashion retail is a dynamic and competitive market and product competition has never been greater. The difficulty for retailers in anticipating and translating trends for their customers is challenging and as many retailers now offer an element of fast fashion. Homogeneity has reappeared in the market and retailers need to seek new ideas in order to attract customers to their stores. Furthermore, a key issue for fashion retailers is product quantity with correct stock in the shop. In terms of product development, retailers are now heavily involved in specifying product requirements as they have to protect their brand reputation, especially if the brand positioning is linked to a specific level of quality or design input. Product is critically linked to brand perception, the quality of a product also ties to its price and is therefore key in ensuring sales and profit.

13.1.4   Retailer branding

Retail branding is an under-researched area and differs in many ways to traditional product branding. A retail brand identifies the goods and services of a retailer and differentiates them from competitors. This leads to the creation of retailer image in the consumer's minds and hence leads to increasing brand equity. Retail brand image is also affected by the manufacturer brands that are sold in store, and the equity attached to those brands. It is also noted that consumers are becoming brand-aware at a younger age than ever, and young consumers have the ability to influence brand choice, particularly in clothing. Therefore, retailers often seek to attract younger customers to their stores with the intention of building a life-long relationship with them.

13.1.5   Retail pricing

Where homogeneity is commonplace, price often becomes the critical factor in a purchase. Retailers should assess the impact of price and price changes on brand image and ultimately the financial performance of the business. There has been little change in the trend of consumers wanting to pay less, therefore the number of promotional campaigns, price reductions, sales and discount days are prominent in the fashion sector. Retailers need to examine the impact of promotional activity across the year and especially during the critical trading period such as double 11 shopping festival, attempting to attract customers into store and encourage spending. The main aim is to achieve volume of sales, rather than profit. This is vital in terms of protecting market share as fashion is easily replicated and a similar item can often be bought elsewhere at a lower price.

13.1.6   Retail service

Retail service is a vital part of the retail mix. When staff members are well-trained and understand the consumer, the shopping experience becomes more enjoyable for the customer and more profitable for the retailer. It has been proven that staff interaction with customers, alongside the physical appearance of store personnel can enhance the shopping experience. Academic research has demonstrated that sales personnel are critical to the store experience and indeed these factors also help customers to decide whether they will return to the store. Retailers must be aware that poor sales staff often causes poor product perception in the minds of the consumer.

13.1.7   The selling environment

The importance of store environment is evident in the literature, and there are a number of elements which create the environment specific to each retailer. Fashion retailers have made a significant investment in both visual merchandising and overall store design to enhance the shopping experience and ultimately encourage customers to spend time in store and purchase at the end of their visit. This is important as the positive impact of store environment is often extended visit times and increased purchase chance. Due to economic constraints, most retailers will adhere to a standard style of store layout where the individuality for the store comes from branding schemes and visual merchandising, which is discussed further later.

13.1.8　Retail promotion

The use of the term "promotions mix" has been common since the 1970s and has since been adopted by academics and practitioners to include a large range of varying advertising and promotional tools. The term is interchangeable with "Marketing communications mix (Marcomms mix)", which describes more-clearly the role of promotion. In recent years, the components of the Marcomms mix have increased further due to changes in technology and the opportunity that the internet in particular has opened up for consumers and businesses.

13.2　Visual merchandising

13.2.1　What is visual merchandising?

Visual merchandising is defined as "the activity which coordinates effective merchandise selection with effective merchandise display". Visual merchandising in fashion is a powerful tool and alongside store design, layout plays a large role in actually attracting customers into store.

Today's market is the "beauty value" battlefield for visual customers. When the post-90s or even post-00s start to dominate the market, the industry needs to change and think more about how to be closer to the consumption habits and aesthetic standards of the younger generation. The first "beauty value" of a brand is often displayed through its shop decoration and display. Only by attracting customers can we give them the chance to understand the brand. Today's market is no longer dominated by products. More and more attention will be paid to all the ways to cooperate with them to enhance "image" and positioning. The purpose of visual marketing is to maximize the potential of the brand and improve the value of goods. The ultimate visual display design can bring consumers a new experience and better grasp the heart of consumers.

13.2.2　Decoration strategies

Customers like convenient and stress-free shopping environment and simple and easy shopping process. Whether the route is smooth or not, and whether the width of the route is appropriate will directly affect the purchase desire of customers. Effective space utilization, that is, terminal layout, should follow the principle of making full use of space. Here are some considerations of decoration.

Eye–catching door and window

The eye-catching door and window can let everyone know the product style and characteristics of the store at a glance. It is very helpful to improve the rate of entering the store. If the clothes you sell are more mature, you should pay more attention to solemnity, temperament and elegance in the display window. If you prefer sweetness, you should focus on sunshine, loveliness, age reduction, etc. (Figure 13.1).

Figure 13.1   Eye-catching door and window

Clean and bright store display

The clothing store should display and arrange hard to achieve novelty and beauty. Moreover, the shape and placement of the shelves should be customized according to the area and shape of the store, and the goods should be hung much but not crowded. It is also necessary to be able to display the featured products according to the features. The featured products should be placed in a prominent position, so that consumers can see them at the first time and attract them to try on and finally buy (Figure 13.2).

Figure 13.2   Clean and bright display

Ornament

When decorating your women's clothing store, we can also add some decorations, which will make the style of the whole store to a higher level and have a strong taste. But in choosing these decorations, you should choose them according to the decoration style of your store (Figure 13.3).

Figure 13.3　Special ornament

Bright fitting room and thin mirror

Some clothes are tried onin the store and look very nice. That's because the mirror in the fitting room is magic that can make customers seem thin and beautiful. Therefore in the fitting room, the mirror is very important. A good mirror will make your performance rise.

Lighting

Lighting design can improve the aesthetic value of clothing stores, and can play a role in changing the sense of space and giving space personality. Therefore, lighting is one of the important tools for the atmosphere design of clothing stores. The main purpose of basic lighting is to extend the light in the whole shop and keep the color in the shop uniform.

Color combination

Although the shape, fabric and structure of clothing also play an important role in the overall combination of clothing, the color factor is better in the sense of people's visual perception and in the psychological emotion of consumers. The color arrangement is especially suitable for

Figure 13.4　Color combination

clothes with basically the same style and rich colors. If you want to achieve exaggerated and eye-catching color combination display effect, the best choice is undoubtedly to display by contrast color combination. It is worth noting that the combined display of colors does not exist alone. It needs to be matched with other factors such as style, shape, fabric, etc. (Figure 13.4).

13.2.3 Clothing display

Relevant data show that scientific and professional product display can promote 30%-40% of sales growth, and a good product display can promote 70% of non-planned consumers' purchase behavior. Everyone will notice this phenomenon: the products that are well displayed and can be easily obtained by retailers must be the products that they mainly promote.

The principles for the display of brand clothing

According to the current market situation, the following principles for the display of brand clothing should be followed:

- Emphasis on humanized design and customer-oriented;
- Highlight the cultural connotation of clothing brand;
- Emphasize the personalized style of clothing brand and avoid the similarity with similar competitors;
- Highlight the selling points of clothing;
- Highlight more profitable commodities;
- Pay attention to artistic techniques.

Commodity display method

Three main display methods are described in Figure 13.5. The advantages and requirements are listed. Figure 13.6 shows an example of repeat display in color.

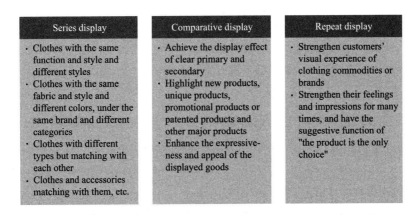

Series display	Comparative display	Repeat display
· Clothes with the same function and style and different styles · Clothes with the same fabric and style and different colors, under the same brand and different categories · Clothes with different types but matching with each other · Clothes and accessories matching with them, etc.	· Achieve the display effect of clear primary and secondary · Highlight new products, unique products, promotional products or patented products and other major products · Enhance the expressiveness and appeal of the displayed goods	· Strengthen customers' visual experience of clothing commodities or brands · Strengthen their feelings and impressions for many times, and have the suggestive function of "the product is the only choice"

Figure 13.5 Commodity display methods

13.3　Sales management

13.3.1　Strategic planning of clothing inventory

How to deal with clothing inventory quickly and efficiently? The marketing plan thinks that dealing with inventory should be "fast, ruthless and accurate", although it is only a short three words, there are many mysteries. At present, the main methods to deal with clothing inventory are as follows:

- Set up a special store or a special selling place for special products in a large shopping mall;
- Deliver commodities in several stages to keep sense of freshness according to the characteristics of products;
- The manufacturer can give the inventory to the dealer as promotional gifts, which not

Figure 13.6　Repeat display in color

only digests clothing inventory, but also serves as a channel reward, which not only increases the loyalty of the dealer, but also urges them to purchase more goods;
- Do not promote sales purely for the purpose of digesting inventory, put the psychological needs of consumers first, and increase sales volume by stimulating consumers' purchase desire and expanding product sales channels;
- Change the circulation channels such as selling in the wholesale market for some old styles at a relatively low price or use group buying strategy;
- Entrepreneurs can donate their indigestible clothing inventory to the poor areas every year.

13.3.2　Clothing style marketing planning

One style combination marketing

In a shop, the structure of each category cannot be divided by simple category structure proportion. Many dealers or brands will set the proportion of each category of the whole set of goods according to their own product advantages. In order to catch every customer entering the store, the brand or dealer must have different styles and types of clothing. The use of clothing style marketing requires the operators to have a clear understanding of the audience of each clothing style, so as to maximize the attraction and meet the needs of consumers through the combination of styles.

Through a variety of style combination marketing, we can achieve a variety of purposes, such as expanding the needs of consumer groups, extending the needs of consumer groups, expanding the needs of target consumer age groups, and developing target customers. Style marketing is not only to attract consumers by increasing styles, but also to make a more scientific planning of fashion styles on the basis of the existing, so as to achieve the maximum benefit with limited cost.

Style update frequency marketing

Every customer's visit to a brand may be accidental, but the frequency of product style update in brand stores will affect the frequency of customers' visit. Setting the update frequency of styles according to the previous sales data is a common approach, which can fully take into account the consumption of commodities in a business district. Style update frequency has a certain guiding role for consumers. The frequency of style update should be set according to the consumer psychological needs of the audience of a certain category, rather than making up for what is sold out. No matter what kind of clothing you are dealing with, you should fully consider the strategy of style update frequency.

Three styles of dynamic marketing

Dynamic style marketing refers to the display and combination of styles to be promoted to customers according to the daily sales situation. For example, the planning of styles and categories combined with the moving line of customers, such as the category connection planning combined with the fashion clothing style, such as the planning of the phenomenon of new goods arrival with the old ones, etc.

Clothing color matching marketing

Color marketing plays an important and appropriate role in a marketing promotion, with proper color and strong marketing force. Color marketing, on the basis of understanding and analyzing consumers' psychology, is to do what consumers think, to properly position commodities, and then to match appropriate colors to products themselves, product packaging, personnel clothing, environmental settings, store decoration and shopping bags, so as to make commodities highly emotional, become a bridge to communicate with consumers, convey the idea of goods to consumers, improve the efficiency of marketing, and reduce the cost of marketing. The application of color marketing in the marketing activities of an enterprise will become more and more frequent, and gradually become an important means for an enterprise to gain a competitive advantage in the fierce competition.

13.4   PEST analysis of fashion retailing

The fashion retail industry is affected by the wide and varied environmental factors including: macro-force, such as economic and socio-culture; and micro-force, such as customers and competitors. The macro-force is generally considered to be made up of political/legal, environmental, socio-cultural and technological factors, and is often referred to using the acronym PEST (Figure 13.7).

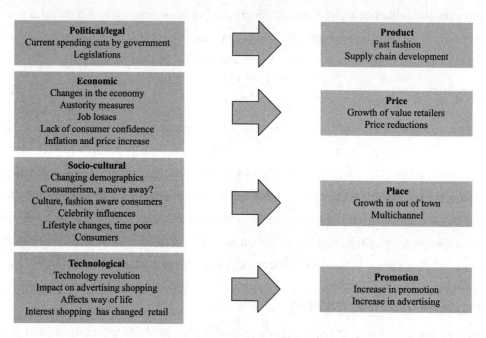

Figure 13.7 PEST framework and impact on the retail marketing mix

As shown in Figure 13.7, PEST framework illustrates the wider external issues that retailers need to manage and understand, and demonstrates their impact upon the marketing mix. Some external factors are common to the whole of the market sector, such as political and economic; however other factors are related closely to the various consumer groups who exhibit different behavior and consequently require varying levels of attention. For example the impact of technology varies across demographics and therefore the incorporation of technology into the marketing mix has to be considered in different ways to meet various consumer needs and expectations.

◎ Summary

Fashion retailing is a distribution channel to realize economic value of fashion design through garment purchasing process. Fashion retail marketing mix consists of a number of specific elements rather than the general 4Ps. Visual merchandising in fashion is a powerful tool and alongside store design, layout plays a large role in actually attracting customers into store. The macro-force affecting fashion retail industry includes political/legal, environmental, socio-cultural and technological factors (PEST).

◎ **Key terms**

Brand loyalty 品牌忠诚度   Fashion retailing 服装零售

Commodity display method 商品陈列方法   Ornament 装饰

Decoration strategy 装饰技巧   Retail marketing mix 零售营销策略

Fashion awareness 时尚意识   Visual merchandising 视觉营销

◎ **Product development team members**

The following product development team members were introduced in this chapter:

Allocator（配货师）: person who is responsible for planning and managing merchandise deliveries received from vendors, as ordered by buyers, to the retail locations. The responsibilities include: arranging for transportation of merchandise to the retail outlet locations; keeping items in stock in the stores and the warehouse; analyzing the needs of each individual store.

Promotion director（促销主管）: person who guides the marketing activities of the retail store.

Retail store manager（店长）: person who oversees all of the activities of a retail store's operation.

Visual merchandiser（视觉营销师）: person who designs, develops, procures, and installs merchandise displays that enhance the ambiance of the environment in which the displays are shown.

◎ **Review or discussion questions**

1. Compare the drawbacks and advantages of garment display in a shop in a specific condition.

2. Imagining you are a manager of a retail shop, what strategies do you use to improve sales in double 11 shopping activity?

◎ **Semester project**

Semester project VII : Develop a floor plan

A floor plan is a scaled diagram of a room or building viewed from above. Suppose you are going to open your own fashion store. How would you make all the perfect layouts? Use the material to make a model out of it.

Here are some suggestions:

1. Use the right type of retail floor plan for your store.

2. Plan your layout based on how shoppers move and behave.

3. Map your store.

4. Keep your decompression zone clear.

5. Put a healthy amount of space between merchandise and fixtures.

6. Go beyond your merchandise…think about the experience!

◎ References

MCCORMICK H, CARTWRIGHT J, PERRY P, et al. Fashion retailing: past, present and future[J]. Textile Progress, 2014, 46 (3): 227-321.

◎ Useful websites

[1] https://www.wipo.int/ipadvantage/en/details.jsp?id=11644.

[2] https://www.alibaba.com/showroom/visual-merchandising.html.

[3] https://www.shopify.com/retail/visual-merchandising.

[4] https://www.retaildoc.com/blog/10-insights-merchandising.

[5] https://www.bilibili.com/video/BV1dL411E7Kf.